Complex Earth

Anomalies Answered

By Steve Preston

2nd Edition

Over 160 illustrations and graphs

Table of Contents

Contents

What You Think You Know?

This book was originally named the Truth About the Earth, but it's more than that so the name had to be changed as updates have been introduced. The most important element of this book is to RE-teach things that should have been presented in our schools or by scientific and religious leaders. You may think you have a pretty good handle on how the earth is the way it is, but after reading this book, hopefully, you will see that you have not been told the truth or at the very least, information you needed to make a reasonable attempt at the truth has not been provided to you. In this edition of this book I have even retimed the Earth events to go along with the new dating methods that are still being taught today. Let's first start by seeing what you know.

How did the universe get here? *- You probably believe that there was a Big Bang and everything started or you believe that God said. "Let there be a universe!", and it started. Neither tells the whole story. There is a third more accurate answer. By the way did you know that they have determined by red shift that the Earth is in the center of our universe and the Big Bang occurred where our earth is?*

How did the Earth get in its position in the Universe? *You probably were told that Bode's law dictated that the earth should be in its position and so it was. That is so very wrong. Bode's "LAW" should be struck from the text books.*

How did animals get here? *You might believe that an amoeba mutated and a "survival of the fittest" evolution started everything to change until we are where we are. If not, you might believe that God made the animals a mere 6 thousand years ago or you believe that God created the animals over a very long time period. A 4th method makes more sense and combines science, history, physical* *evidence, and religion rather than requiring one or the other to suffer.*

How did the dinosaurs get so huge? *You possibly are thinking survival of the fittest again or God zapping the huge monsters into existence again. Let me make it perfectly clear, the Bible says no such thing and the survival of the fittest hypothesis would not have resulted in dinosaurs. While neither of these is the answer, there is a reasonable answer to this question.*

How did the huge mountain ranges develop on earth? *You probably believe that plate tectonic action smashed pieces of the earth together and there they were. You will find that that answer many times is absurd. A much more reasonable answer awaits you in the book.*

How did the Pacific Ocean get here? *You might believe that the earth was formed that way or believe that some massive meteor struck the earth. Neither appears to be correct. The correct answer has been well studied and requires one of the planets?*

How did Mars get split in half and what does that have to do with the Earth? *You might be saying, "If Mars had been torn in half, you would have heard about it before now." Unfortunately the answer is you SHOULD HAVE been told about it.*

How did the dinosaurs become extinct? *It was not rats eating dinosaur eggs or a massive layer of iridium dust. There is actually a logical reason.*

How and when did Venus turn into a ball of fire and how did that affect us? *- If you believe that a thing we call the Greenhouse Affect affected Venus---You are wrong. If you believe that Venus caught on fire millions of years ago you are wrong again.*

When did the Huge split that almost splits Venus in half occur- *You probably are thinking, "There is no such split!" While no one has told you about it, the huge split is important to earth's history.*

How did the Mammoths and other animals get quick frozen 10 thousand years ago? *If you believe an Ice Age did it, I'm sorry to say that is impossible. If you believe that mammoths lived in the arctic and just happened to freeze with temperate zone flowers in* *their mouths, you are wrong again. If you think they died as a result of the worldwide flood of Noah, you have a third strike. The answer is confirmed by many repeats of the same action and will be fully explained.*

Were Neanderthal humans dumb? *If you believe that we are smarter than Neanderthal humans, try to explain why their brains were larger. The explanation is simple.*

When did humans become civilized? *If you believe they became civilized with the beginning of the Bronze Age, that's not far enough back. If Neanderthals is your answer, you are still way off. If you believe that Adam was the first civilized human, again you are wrong and you have not read the Bible Genesis story completely. While the evidence is overwhelming and convincing, some will have a hard time with this major factor in earth's development.*

Was there a worldwide flood and did Noah carry 3 million animal types in his boat to safety? *If you believe that Noah brought all the animals to safety in his boat like many do, you are wrong. The kangaroo, for instance never saw the Middle East. While the first statement seems to be different than the Biblical testimony, it is not. If you believe that no flood ever occurred, you are very wrong according to masses of evidence. If you believe that a massive canopy of water burst to flood the earth during Noah's time, the probability is extremely low. If you believe that rain flooded the earth you are wrong as well. I know this sounds different than what you believe the Bible says, but a more appropriate probability is discussed in full and it should open your eyes.*

What made people's life span decrease so dramatically 6 thousand years ago? *If you think it was the worldwide flood, you are wrong. By all evidence the flood occurred 10 thousand years ago and people lived long lives after that fateful occurrence. If you think people never lived longer than we do today, you are probably wrong again. If you think that it has something to do with the Bharata War, you are right, but what do you know about this massive worldwide war?*

When will the earth be affected by global warming and the greenhouse catastrophe? If you believe soon, you are probably wrong again. If you believe the earth will not change in the near future you are wrong again. The answer is confirmed by prediction, scientific study, mathematical modeling, and physical evidence. With almost certainty, the earth will not get hot. The answer is scary and it is soon.

If you're looking for a science book that ignores the more unpleasant things that have been witnessed either as physical evidence or by first hand descriptions and knowledge, you had better just shut the book right now. If you are looking for a book that will comfort you, you also have come to the wrong place. If you really want to learn about how the earth got to be the way it is today regardless of what you were told in the past, please continue.

Continent of Prestonia

During the Permian Age, life was great with 2 great continents Pangea and Prestonia, but at the end of this age, the current theory is Mars came too close and ripped away the entire continent of Prestonia leaving a huge hole called the Pacific Ocean. It was the beginning of the Mesozoic Era. The graphic below shows the results. Luckily no people lived on the planet at the time but that soon would change. The event is not only recorded in the Antarctica Ice cores, but in many other pieces of evidence.

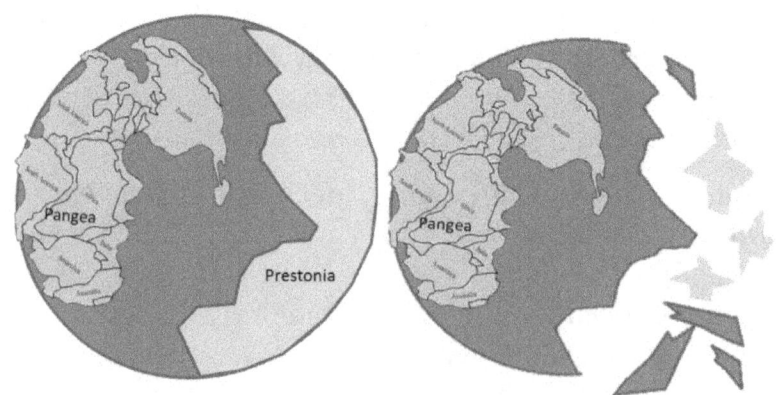

The First People

During the Mesozoic Period, people existed and lived with the Dinosaurs. Great societies arose. Sciences flourished. The people were called the "Great Men of Old" by the Jews and "the Titans" by the early Greek. Evidence is found around the world of Mesozoic people up to 12 feet tall, Mesozoic batteries, Mesozoic nuclear plants and all the things that show a vast civilization. There is no question that some were on one side of Pangea while some portion went across the Atlantic River to become the first Americans. The continent slowly separated but life flourished on both Eurasia and America all during the Jurassic and Cretaceous Periods as shown.

These Titan people vanished along with the dinosaurs at the end of the Cretaceous Period to be replaced by the people called the Nephilim, ANAK people [Jew], Anunnaki Sumeria], Olympians [Greek], Lords of Amenti [Khemetian], the Araya [Persia], the Archaics [North American], or the Akamim people [South American] depending on whose history you are reading at the time. The ANAK were rulers of the world as the Cro-Magnon came on board about 40 thousand years ago. The ANAK had control of both Eurasia and America. There were bitter wars without a true winner as America drifted farther away as shown below.

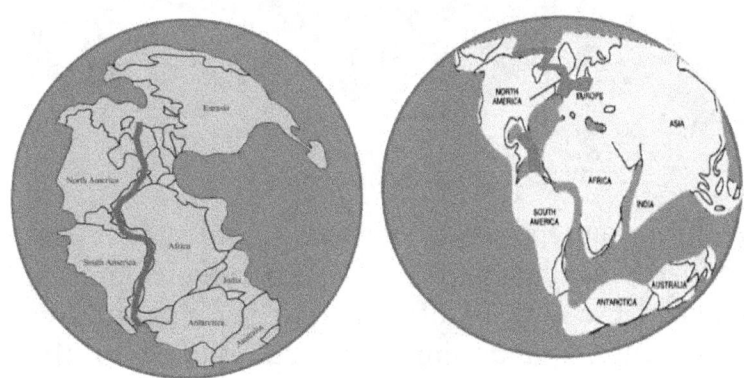

Venusian Moon

We are not at the end of the Pleistocene, but flying became normal for trade and war. Apparently, trade between Venus and the Earth was ongoing and a group of colonists were resident during massive warring. Called the Rahab, vain planet, Venus was destroyed as the electrical differences between Earth and our close neighbor became too great. The unfortunate position of the Venusian moon caused it to shatter. This peppered the Earth with hundreds of thousands of

9

Meteors making craters everywhere that are still extant along the East Coast of the United States. We don't know how many hit the earth, but there are still hundreds of thousands of craters left from the massive onslaught. The earth shifted on its axis, massive weather flooded the earth, and millions of people and animals died. On Venus, it was worse. Hundreds of thousands of moon pieces hit and disrupted the planet rotation so much that it was set on fire and it is still burning today over 10 thousand years later. All animals and people on the planet were almost instantly killed, cities were melted, volcanoes erupted around the planet almost simultaneously, and the water quickly evaporated. Venus became a dead planet as shown below right. To the left is a tiny section of the meteorite field with 42 of the thousands located along the coastline of the United States. They are called the Carolina Bays, but during the impacts, massive fires raged out of control.

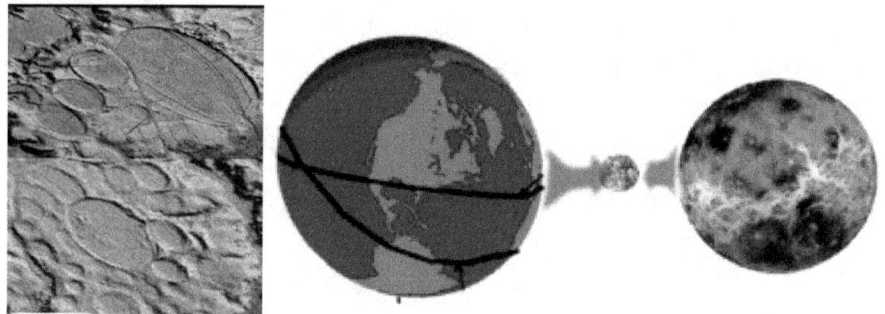

A New Stone Age

After the Earth settled down, societies flourished again as the people living during this time had flying vehicles and ships that circumnavigated the earth. Unfortunately, six thousand years ago another war erupted and almost wiped out civilizations. The Nephilim people lost control and the Cro-Magnon people who survived were sent back to a quasi-bronze Age. People were spread all over the world according to the Biblical texts and many other ancient writings. Soon communication between nations vanished.

Bizarre Is Not So Bizarre

Because people are not typically told about less than comfortable physical evidence, some things seem bizarre to us when we read about them. Hundreds of pieces of evidence tell us a different story than what you are used to reading. For instance-

- What would you think if I told you that Mars made the Pacific Ocean?

- What would you say is I told you that Dinosaurs weren't heavy in the old days and Titans were actual people?

- What is I told you that Venus used to have a moon that affected the earth?

- What would you think if a shoe print and many others like the one shown next were found incased in stone?

While all of these things seem bizarre, let me examine the shoe a little more to give you more food for thought. Inspection of this shoeprint makes this find even stranger in that it shows that 2 trilobites were crushed by its wearer.

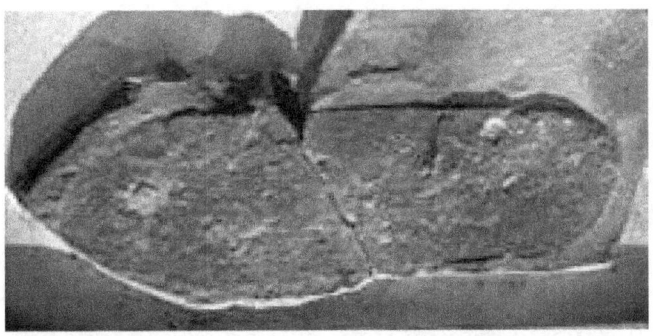

Knowing that trilobites lived during the Mesozoic Period and the fact that it takes a long time to make stone, one would assume that the shoe wearers lived during that same time. Now multiply this find many, many times as more and more evidence comes to light showing that civilized humans walked with dinosaurs. People have found jewelry, articles of machinery, batteries, modern human bones and constructed walls all encased in stone and deep in the earth. The obvious answer that has been adopted by many scientists is shown below.

1. *None of the evidence means anything.*
2. *The evidence is flawed.*
3. *Rock doesn't take long to form.*
4. *Trilobites lived recently.*
5. *The historical records are myths.*
6. *The human bones found are ANOMOULOUS.*
7. *The artifacts are somehow faked.*

This is how we get "comfortable Histories, but it is not how we find the truth.

This one example is backed up with multitudes of supporting evidence and data. The label of ANOMALY is put on much of this evidence as the science and religious communities try to ignore elements that don't fit into their view of existence. Instead of turning away from unimpeachable evidence or using hair-brained theories bent on making you comfortable, you're going to find out about some really interesting and important concepts that will allow you to assimilate the evidence that is all around us. Next you will find a few of the seemingly inappropriate scientific assertions that will begin to make sense as you are presented the evidence that is typically ignored. They are a little different than the "theories" you have been taught. Rather than just accepting what you were taught, be prepared to be amazed at what is not normally told to us as we go through school. You have purposefully been kept in the dark about the world around you, but ignorance will not make truth go away.

- **The Bible and science work together**. I don't mean by twisting them around, but by using their innate truths to support one another.

- **The Big Bang didn't simply happen**. An entity was responsible. What that means will be explained later.

- **Before the Big Bang there were at least 2 universes** that did something scientists now call the Big Splat. The 2 universes are still around according to modern scientists, mathematicians, and religious dogma.

- **The Evolution Theory doesn't work**. Nothing fits. The timing, the number of animals, the rapid expansion of animal life immediately following an extension period, the weird mistakes of nature, the lack or a transition model, the survival of the most unfit and many more curiosities all tell us the same thing. While it doesn't make sense, uncontrolled evolution of species is still taught as fact.

- **God creating all the animals doesn't make sense** either and the Bible doesn't indicate that he did. We can even get an approximation of what animals were created and those that came about another way. Those that were not "created" were deemed "abominations" or "unclean animals" in many ancient texts. I know you were told that unclean meant Jews shouldn't eat them, but it is not the reason they were called that.

- **"E=MC²" doesn't work.** Einstein knew and others know it today, but it is taught as if it were a law.

- **Invisible mass is a new concept for many**, so I will explain what it is briefly and what it means to us. Mathematical string theorists require its existence and so does Religion.

- **Creationist theories don't work.** While I don't go into this in detail, the disregarding of modern dating methodologies to make the claim that the world is only 6 thousand years old is not practical, reasonable, or founded. Oh yes!--The theory is not Biblical either.

- **The continent of Pangea was not "the" super continent**. There used to be more than one. The other massive continent was never named and never discussed in school.

- **The Earth almost destroyed Mars**. That incident made the Pacific Ocean. Mathematical models tell us this, physical evidence tells us this, and ancient historians told us this. Now I'm going to tell you and show you the details.

- **The earth almost "split in two"**. Huge Magma Mountains attest to the fact and the danger. Here's the kicker. The splits have occurred on the opposite side of the world whenever huge craters were formed to define gigantic meteor strikes.

- **The earth has flipped on is rotational axis** many times. When it flips, it does it quickly and animals are quickly frozen or twisted together in huge piles. Surely you have heard about the frozen animals, but have you been told how it was possible to "quick freeze" a massive animal like a Mammoth?

- **Civilized humans lived with the dinosaurs**. That doesn't mean that dinosaurs were living millions of years ago, it means that humans have been here a long time and the proof is amazing. ---No! The humans did not come from outer-space. If we look at esoteric ancient books like the book of "Genesis" in the Bible and many others, we can find the truth that agrees with huge quantities of physical evidence about this fact.

- **There is strong evidence for Atlantis** and the sinking event actually happened. I don't mean by reading a book by Plato either.

- **The moon of Venus exploded 11 thousand years ago**. Yes! - I said its moon. When the explosion occurred, Venus caught on fire and earth was pelted with thousands and thousands of meteorites. Venus itself almost split in two and the evidence is everywhere. We can even date the event very accurately and we can get agreement in ancient religious texts around the world including Jewish texts. All this happened 10 thousand years ago.

- **The Biblical Tower of Babel event actually happened** 6 thousand years ago, but more importantly, strong evidence shows that a nasty germ was let free that decimated the population of the entire world at this same time. The effect of this agent was the MOST important thing affecting the development of civilization from that time forward.

- **The world is not heading for a meltdown** like Venus. Instead, a terrible Ice Age will be here within our lifetime. Scientists have been recording details of this event now for years.

Absurdness Recognition

I know all these things sound absurd right now, but soon I think you will begin to see what has been "hidden" from you. With that being said, let's get started. What better place to start at the beginning. It was the beginning of the beginning so to speak.

1. **Section 1-**The first part of this book will deal with Macrocosmic definitions being used today to support a truer description of how things got here.

2. **Section 2-**The next section will be a description of the beginnings of our solar system and the earth.

3. **Section 3-** Following the overview of earth problems, I will concentrate on the affect Mars had on the Earth.

4. **Section 4-**After the earth and Mars settle down, there is a discussion of the key elements that molded the earth. Meteors, rotational anomalies, volcanoes, and the earth splitting all characterized the major portion of earth's history. Sometimes their details are ignored to make us feel safe.

5. **Section 6-**Evolution is tackled next with surprising evidence and conclusions and the evidence of an ancient civilized human existence on the earth follows.

6. **Section 7-**The effect of the Venusian Moon exploding is the next topic. The evidence will certainly surprise you.

7. **Section 8-**The Worldwide Flood and Pleistocene Extinction identified in the Bible is discussed next. The evidence may open your eyes and bring your religious beliefs you're your scientific understanding closer together.

8. **Section 9-**The Tower of Babel and the Bharata War is the next to last topic, but it is very important as we try to determine why people only use 10 percent of their brains. I know it doesn't seem like a tower has anything to do with the brain, but you will again be surprised.

9.   **<u>Section 10-</u>**The last section is on the upcoming ice age. Get out your winter clothes.

If you are you ready, let's look at the Big Picture.

Timing is Everything

Bang, Bang, Bang, Bang, and Bang---It happened 15 Billion years ago. Some call it the beginning of time, but whatever you call it; many agree that there was a fast expansion of matter that filled large areas of our universe. Before we get to the earth and see its development we have to create the universe to put the earth in. Have you ever wondered how the universe got here? If you have been dooped into believing that the Big Bang Theory satisfied the beginning of the beginning? I'm sorry to say that it does not. I'm not going to get into this subject too much because an entire book could be written around this one event. The only thing I want to do here is to establish a concept. That concept is an important one and a simple one.

You can't separate religion, history, and science. That doesn't mean the universe is not millions of years old, but it does mean we need to look at evidence from multiple sources.

They are not mutually exclusive as many seem to think. It seems that when people are in a religious moment, they think about how the religious world works and when they are not, they kick in this concept of science that is completely free of the "strange" parts of religion. Historians simply write what they would like to have happened regardless of any religious or scientific affect or defect. Somehow, there must be a union of the three or they become useless to us and they all become ineffective lies. Given that we must use science as a way of seeing this world around us, I suggest that we use historical records and religious truths as well. You might wonder how I could say that science was a lie if religion was a lie. Well, I'm not going to tell you right now, but I think that, as you go through this book, you will see what I mean. Our first example is this big explosion. There are many questions that are not answered by the explosion and many anomalies that go against the theory completely. They cannot be ignored and many cannot be "mathed" away nor can someone arbitrarily refute evidence suggesting this major event simply because carbon 14 dating is now known to be unreliable. I know that carbon 14 would not have been used to determine a 15 Billion year old event, but some people keep bringing up that particular test method limitations to justify a completely religious based beginning "story" in a non-science science called creationism. The creationist theory has a beginning of everything starting only 6 thousand years ago which is totally determined by a misinterpretation of Biblical texts. Today those who try to refine what we know about our beginnings don't rely on the Bible by itself or carbon 14. They use many, many techniques to insure that the timing is as accurate as possible.

Dating Problems

Today we know that nuclear decay dating is unreliable and gets absurd after about 50 thousand years. Therefore, other dating methods had to be used to provide a more correct dating. Luckily Ice was packed in the Antarctic and Arctic regions of the earth. By a substantial amount of testing we can now see physical evidence of massive extinction creating cycles in ice cores that are verified by other Ice Core samples, Archeo-Magnetic dating under the Atlantic Ocean, O18 Marine isotope cycles, and Hotspots tracks under the Pacific Ocean. Together they give us a different picture that lessens the anomalous characteristics that are challenging stratographic testing and other methods being use as a standard until today. The following graph shows the major extinctions that show as massive temperature and CO_2 changes around the world and recorded in Ice. These events are correlated with more details

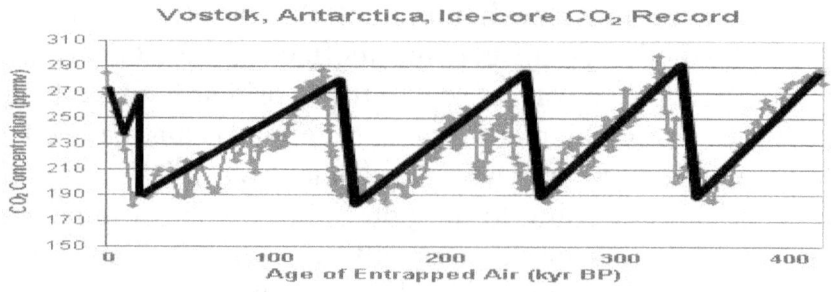

Vostok, Antarctica, Ice-core CO_2 Record

More Detail

I know you are thinking this is interesting, but it doesn't really help too much. So we have correlation of hot spot trails and what seems to be timing compression similar to others used in this new light, but can the hot spots be traced back to the Antarctic Ice core? The answer can be seen in the next graphic. We know that the trails are produced perpendicular to the axis of rotation of the earth which is described as dotted lines below.

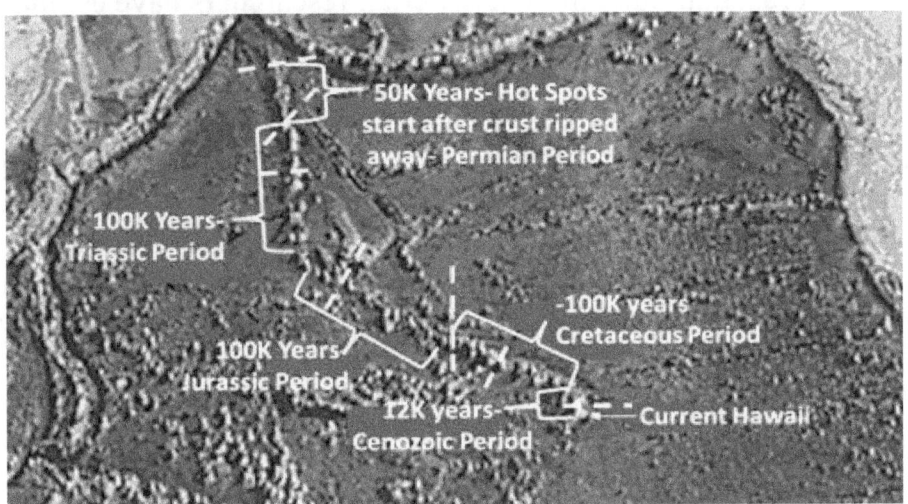

If we that the changes in the earth axis at the apparent changes, we find something VERY interesting as shown next.

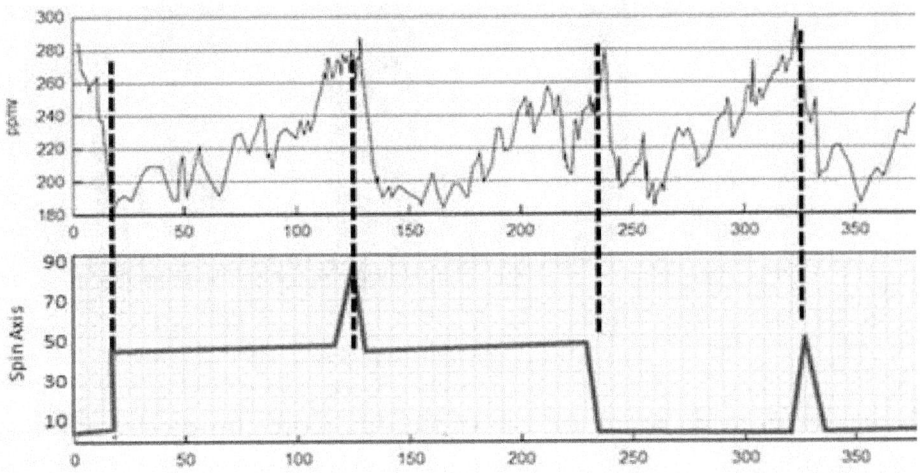

19

With that, let's look at the center of the Atlantic Ocean. The graph following shows the last 14 flips over what had been determined to be a period of 3.7 million years using nuclear decay Potassium-Argon dating of magnetic material in solidified magma in the center of the Atlantic Ocean. Not only is a general time of each flip noted, but also the ferrous portions of the magma align with the magnetic field of the earth to show rotation of the earth over time. Another flip could happen any time. Using mathematical models of the external crust and inner molten material, researchers have estimated with mathematic models that the <u>Earth should flip on its axis about every 100 thousand years</u>. The problem with trying to determine the actual workings of the Earth is that no one has ever seen the inside of the Earth to model it properly, but the results do confirm the high possibility of a polar flip, which will cause mass destruction, tidal waves, and major climatic changes. With that scary introduction, let's look at the chart as it currently has been determined and understand that nuclear decay dating is not nearly as accurate as we once thought.

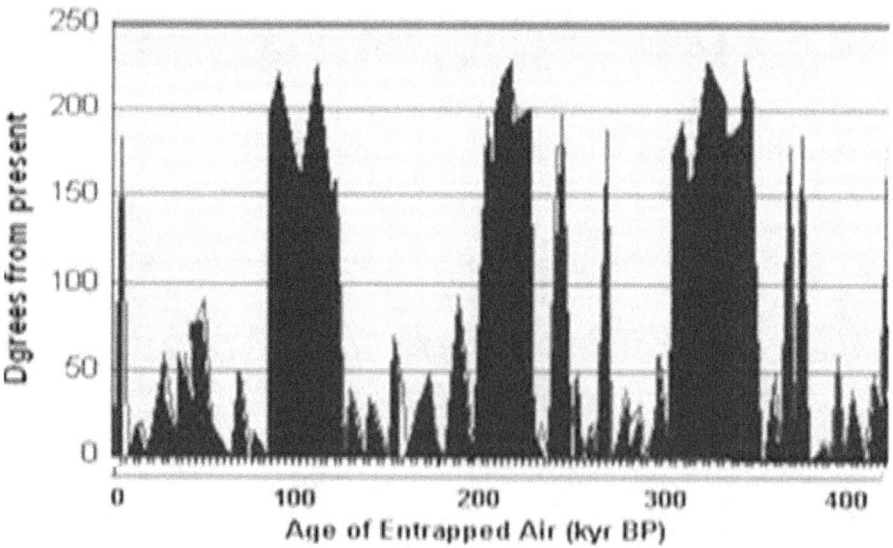

When we put this graph on top of the Ice Core timing we get more verification.

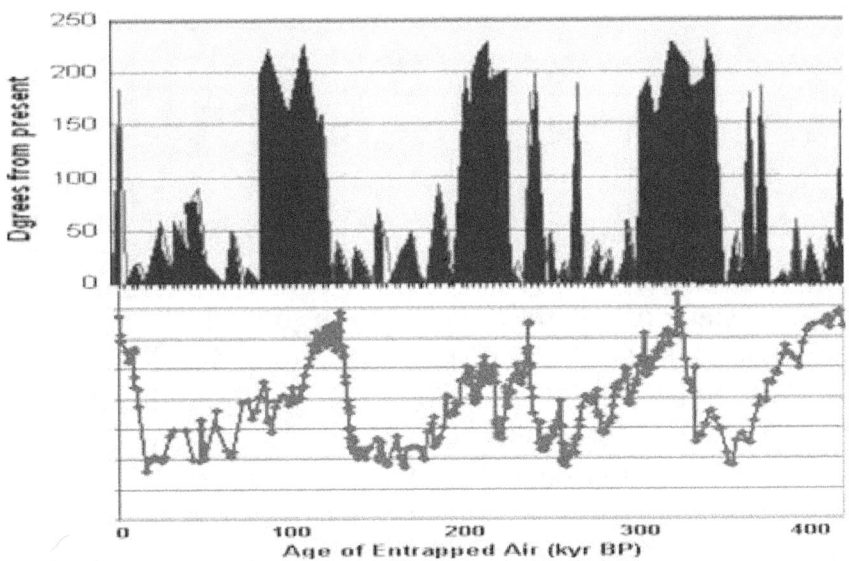

By this it shows massive shifts occurring 10,120, 250, 350 thousand years along with some other bobbles, but look how well it matches the Ice core and hotspot track data. Instead of what you are used to, we find the following.

Era/Period/Epoch	time (T yrs. ago)
Archaeozoic period	50,000-3000
Proterozoic period	3000-1100
Cambrian period	1100-950
Ordovician period	950-850
Silurian period	850-750
Devonian period	750-650
Carboniferous period	650-550
Permian period	550-450
Triassic period [Mars Encounter]	450-350
Jurassic period [Titan]	350-250
Cretaceous period [ANAK]	250-120
Tertiary period [Adam]	120-40
Pleistocene period [Venus End]	40-10
Holocene period [Present]	10-0

Controlled Big Bang Theory

I'm going to start off with a concoction of all the isolated mathematical models of the "creation" including the Membrane "M" theory, the Super-symmetry theory, the Big Splat theory, and of course the Big Bang theory. They all lack conclusive determination. A more reasonable explanation of the beginnings of the universe and the earth is something called the "Intelligent Design Big Bang". There is a lot of evidence to suggest that there was an explosion that occurred billions of years ago alright, but how the universe got that way, what happened to much of the matter and many other errors brought out in the various problem filled theories above can be cleaned up fairly easily if we consider an outside force.

The establishment of an event boundary like the big bang event is a thousand times easier to explain with a controller. This controller would have been responsible for the <u>culmination of compaction</u> that eventually caused the Big Bang reaction. With outside intervention comes a more plausible conclusion and it is neither a cop-out nor a nonscientific description to say that God caused the Big Bang. If you are uncomfortable with that statement let's say God at least, initiated the elements that would come together to establish this important event boundary. While the Big Bang is considered an important beginning for something, according to a lot of math, it was not the beginning of time nor is it satisfactory to simply say God made the universe. I'm going to get a little weird here, but it will make sense in a bit. What we might want to say is God made VIBRATION.

Vibration and Our World

I know vibration sounds like a mundane component and to relegate God into this vibration component seems to be saying God was not a major contributor to designing our universe, but you would be wrong. Only very recently have we even been able to approach "God's Vibration" to develop questions.

- **Many recent studies** have forced us to throw away Atomic theory and look at everything with something called the vibrational matter which states matter is really vibrating Aether than has no mass until it vibrates.
- **The Theories of Dimensional Reality** have been shattered with the new string theories which rely on an entropic component that is defined by its <u>vibration</u>. [I know that sounds like big words to make me sound important, but I think you will see what I'm talking about pretty soon.]
- **For instance**, we now know from experimentation and mathematics that this fundamental <u>vibration</u> of particles makes things larger, smaller, **<u>invisible</u>**, weightless, heavy, or visible.
- **Vibration determines what particles** will or can come in contact with others.
- **Vibration establishes the entropy** variations that allowed clumped up galaxies to be formed in the universe.
- **Vibration is also the element** that establishes the elusive GRAVITY component of masses.
- **Vibration is the secret** to understanding all matter.

Here's the kicker. No one can explain how the vibration was made or exactly what it is. The other thing that is mathematically improbable is that everything is completely tied to this vibrational component. It is something called a mathematical single point which always involves a manipulator. I'm sure this sounds strange to you and it should. Remember, I'm talking about a single particle that makes EVERYTHING—and here's the bizarre part; this includes everything that is invisible. We are also talking a single dimensional string, like the dimension of time the dimension of length, etc. and how they can be manipulated by changing its fundamental "VIBRATION". Let me give you an accidental example of what happens when people disrupt only the vibration component of matter. For that let me introduce the Canadian scientist named John Hutchison.

John Hutchison Experiments

John has been experimenting with vibrations associated with Microwaves. These are the same microwaves that cook food. By bombarding materials with a number of different vibrational frequencies, he has done some well documented and amazing things like making heavy objects float or making one object invisible to another or turning metals into a pudding-like substance without heat. Below left is a picture of metal whose particle vibrations were changed for an instant. With its vibrations modified, it was no longer metal and it wiggled like Jell-O. When the vibration disruption was stopped, the material again became metallic.

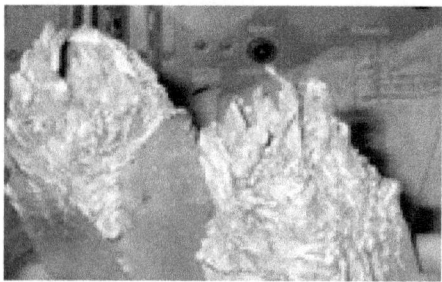 

Above right is a bowling ball that became weightless when bombarded with ultrahigh frequency waves. To say that John or any of the other researchers know what makes the vibration would be a reach, they are simply using God's vibration component to modify matter. People can modify vibration or split it into two or more parts, it cannot be created.

GOD Created Vibration.

Noise Canceling Example of Invisibility

While I'm at it, let me give you another example. Noise canceling headphones work on a simple process. If you receive sound in both ears with the same phase and frequency it becomes invisible to you. You cannot hear it. Use that same analogy and you will see the beginnings of matter. If two of these tiny vibrating Aether masses come in contact with one another and they have the same vibration, when they are together, they become invisible to others that are attuned to that frequency. If particles are NOT invisible they can combine with one another and become gigantic atoms. Each atom has a frequency component established by its Aetheric parts.—Change the frequency and the Atom changes—i.e. lead could turn into gold.

Equation for Invisible Mass

Here is where the universal Mass equation comes in. $E=MC^2$ is generally known, but ALL energy equations are of the for $E=\frac{1}{2}XY^2$ Potential energy, Inductive energy, Capacitive energy and all the rest follow this same universal equation constant. The true Mass-energy equation must be $E=\frac{1}{2}MC^2$. By this revelation and understanding the visible mass operates to the $E=MC^2$ equation we know that $\frac{1}{2}$ the mass in the universe is invisible. From this it may be easily understood that the invisible matter is visible somewhere else. Pretty soon the idea of parallel universes, like the universe called heaven, can coexist in the same location without any major interaction. The establishment of how a heaven [parallel universe] MUST exist, according to modern super-symmetry and membrane theories is now bringing science and religion closer. The vibration component of mass also defines how levitation, invisibility, and other previously religious concepts are not religious at all. They are REALITY concepts.

In the Macrocosm, this vibration establishes a method for continuous change with a "requirement" of repeated history. God is the designer and caretaker of that as well. Its beginning is certainly controlled by God as its end and/or rebirth is controlled by a vibration component. It's like saying, *"We have many equations that define how vibration establishes everything, but without an intelligent controller, the vibrations have no beginning."* I know that was a roundabout way of defining "intelligent design Big Bang", but like is said, in this first section, I really wanted to establish the truth that religion and science are not separate and "the beginning" cannot be adequately defined. I also need to state that two items have NO mathematical beginning.

God Created Vibration---the Universe came into existence when vibration was created.

God Created Life---Certainly life on earth began at this time.

Chromosomic Life

[Pretty neat word isn't it?] If you remember how we were taught that various chemicals sat around for millions of years and finally they came together in a specific way to produce life; it sounded like a marshmallow definition to me!!! There was no "cause" in the answer except for the extremely long time periods identified.

Now Let's Look A Little Deeper

Everything that has life is chromosome based. Isn't that odd? The structure of life is IDENTICAL for all living things. Doesn't it seem that after millions of years of randomness, different structures of life would appear? Just like I mentioned concerning vibration, chromosome based life as a single point is not mathematically probable. It is the only thing that can make life and we have no idea how it got here because it does not fit into a mathematically controlled universe, it was CREATED by God.

We will get into the "LIFE" part a little later. Right now we need to examine the birth of our solar system. The Solar system came about in a simple and explainable way. Unfortunately, like the big bang, some of the solar system's developmental elements have not been presented in our schools and our children need to know some of these things or NOTHING will make sense. They won't understand our EARTH.

Solar System Was Born

We are told that the solar system generally was established a long time ago and with it, the planet Earth "sort of" came into being. New evidence strongly suggests that there was more to it than a simple spinning group of clumped gasses that were forced into orbits determined by some weird "Bode's Law". It's looking more and more like many of the planets did <u>generally</u> assume their place, but many years ago a planet close to Mars was orbiting and no Earth was evident 93 million miles from the sun. The diagram below shows this first Solar System makeup. The large planet next to Mars becomes earth, but right now it is simply the planet next to Mars and it is a big one. The large mass at the farthest point from the sun is the Oort cloud that is still extant.

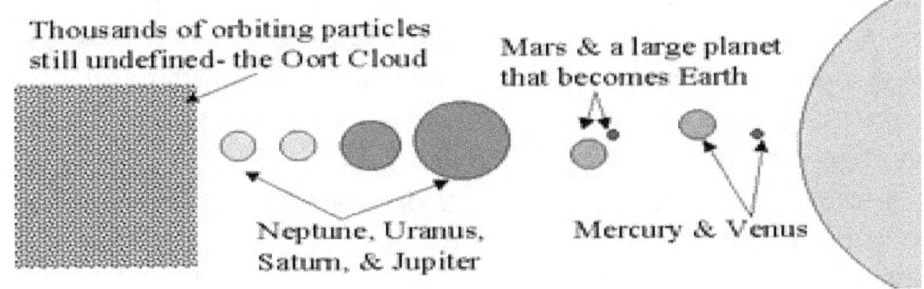

Thousands of orbiting particles still undefined- the Oort Cloud

Mars & a large planet that becomes Earth

Neptune, Uranus, Saturn, & Jupiter

Mercury & Venus

The Solar System Changed

Current evidence tells us that the Solar System changed a mere ½ million years ago. The planet close to Mars finally exploded from the gravitational pressures of the 2 close planets and the asteroid belt was established. The largest part became the Earth. I know that concept seems fanciful but please continue reading and its probability will become more understood.

Beyond not believing this event occurred ourselves, we have a very difficult time believing that this event would be understood by ancient man. Around the world the same story was told in ancient times. A super planet was split in two and the earth was made along with something called the "heavens" or the "firmament".

New computer simulations of this event tell us the same thing. One probability concerning the asteroids is that they were made from some fly-by-night rogue planet, the earth and Mars almost collided and the result was what we have today. You may not believe this fact right now, but the proof is provided in later chapters specifically dealing with this critical time period for the earth.

Similar Stories

I've put together 17 substantially similar creation stories from around the world. Everyone told the same story. From one culture to another, the details are so close that it's almost impossible to ignore that they are, basically, the same story. Their similarity is not dependent on closeness of culture or placement of the people. In fact, a common theme in these historical references is the concept of the Earth splitting in two. Half became what we call the Earth and a firmament was made from the other half of the planet. What on Earth would have given the ancients such a strange concept? From these excerpts one can begin to get an understanding of what might actually have happened or what a large segment of the people believed happened, millions of years before the time of the stories being told. Sumerian, Aztec, Egyptian, Chaldean, Hindu, Ute, Maori, Celtic, Hopi, Chinese, and Polynesian people have no basic history together, but they all have the same insight and somehow had knowledge of what must have happened when the Earth was formed. I've tried to keep the overviews short so that the basic concepts and similarities can be quickly seen. Afterwards I have presented a couple of interpretations to consider given these similar stories.

Don't throw data away just because it is strange. When trying to examine the past. USE the stories of the past!

Note: the bracketed information is mine and just identifies similarities a little.

Chinese Version

[From Chinese legends]

*The first living thing was P'an Ku. **[God]**. He evolved inside a gigantic cosmic egg, which contained all elements of the universe. The sky and Earth were one. **[Sky and Earth was a super planet]** He separated the Sky from the Earth. Gradually he separated wet and dry, then the light and dark. While he grew he also created the first humans.*

Hindu Version

*Brahama was the progenitor of the whole world **[The "whole world" was the "heaven/Earth" super planet]**. After a while, the divine one, by his thought, divided it into two halves. **[Earth split]** From the two he formed Earth and the firmament **[asteroids]**. The eternal abode of the waters was made, and then he created all beings.*

Aztec Version

*Originally the Earth and Sky were one named Coatlicue. **[Coatlicue was the "Earth/sky" super planet]** The creator ripped her into two pieces forming the sky and the Earth. After the creation, the gods saw that the Earth was formless. Only the ocean was everywhere.*

Egyptian Version

[From the Egyptian Book of the Dead]

*The God Re-created everything including Ra, the sun god. Then Ra got himself pregnant 1/4 of his offspring became the Earth and three fourths became the sky. **[Ra's offspring combination was the "Sky/Earth" super planet]** Light streamed forth, banishing darkness on the third day.*

Digueno Tribe Version [Africa]

*In the beginning was Tu-chai-pai. He made the world. The male part was the heavens and the female part was the Earth. **[The male and female part was the "heaven/Earth" super planet]** Tu-chai-pai said "We-hicht" three times, which caused the heavens to rise above the Earth. **[Earth split]** First he made the ocean and the land; He then made man out of mud.*

Nkongolo Tribe version [Africa]

The Earth and heaven were created together. *[**This was the "heaven/Earth" super planet**]* The Earth and heavens were separated. *[**Earth split**]*

Thailand/ Laos Version

*In the beginning were "the three Great Men" [**trinity**] the heavens and the Earth were joined together by a rattan bridge [**The joined heaven and Earth was the "heaven/Earth" super planet**] the great men placed the Thens [**angels**] to rule the heavens. The people refused to worship the Thens. The Thens overwhelmed the world with a flood. Some of the Thens came from heaven to teach people skills including metalworking, music, and cultivation. When the Thens returned to the heavens, the bridge between heaven and the Earth was destroyed and all contact between the two ceased. [**The super Planet Split**]*

Polynesian Version

*The supreme God Io, created the world [**This world was the "Earth/sky" super planet**]. In the beginning there was only waters and darkness. By his word and thought, Io separated the world into the Earth and sky. [**Earth split**]*

Maori Version [New Zealand]

*Tane the Creator God made the Earth and sky as one [**This was the "sky/Earth" super planet**]. The world was full of darkness. Tane forced the sky away from the Earth. [**Earth split**]*

Celtic Version

*God made the Heaven and Earth as one [**This was the "heaven/Earth" super planet**]. Titans were on the World. One of the Titans cut them apart and split the heaven part into many pieces. [**Earth split and asteroids were made**] This allowed light to shine on the Earth.*

Northern Europe Version-

[From "Poetic Edda"]

*First there was a huge giant named Ymir. [**Ymir was the "Earth/heaven" super planet**]. The trinity of Gods killed Ymir and* *he was split in two. His body became the Earth and the head became the sky.*

The Hopi Indians-

[From the "Book of the Hopi"]

*The creating power made the world before this one **[This original world was the "Earth/sky" super planet]**. He was displeased with the people. He stomped on the world. It split in two [**Earth split**].*

Snohomish Indian *[Tradition]*

*The creator deity named Dohkwibuhch, created the world. Dissatisfied with the world, a wise man discovered that everyone bumped their heads against the sky, as it was very low. [**The earth and sky were originally together**.] Some people climbed up high trees and went into the sky world. [**There was communication between heaven and earth in ancient times.**]The wise man said that if all the people and all the animals and all the birds pushed at once, the sky could be lifted away from the Earth and it was so. **[The earth and sky were separated.]***

The Ute Indian Version

*[**See how very similar the Ute Indian and Sumerian versions are.**]*

*The sun god, Ta-wat, and the god, Tavi, were in the beginning. **[Tavi was the Earth/heaven super planet]** Tavi came so close to Ta-wat. [**War in Heaven**] Ta-wat shot an arrow at Tavi's face. [**God captured the rebel angels and the created creatures**] Tavi was slivered into 1,000 pieces **[Earth split and Asteroids were made]**. All that remained was Tavi's head **[Tavi's head was the Earth]**.*

Sumerian/ Babylonian Version

*In the Sumerian version Omorka was called Tiamat, but most of it was identical. The gods Marduk **[an outside planet]** and Omorka were created. **[Omorka was the Earth/firmament super planet-]** Omorka ruled over the "men with wings" **[angels]** and beasts*

31

created by chaos. Marduk came too close to Omorka and disturbed Omorka. **[There was war in heaven]** Marduk He sent an arrow into Omorka's belly it split Omorka in two. **[The Earth split]** The half of her he set as a screen **[the firmament {asteroid belt} was formed].** Omorka's skull was cut loose. **[A new Earth was formed which went to a new orbit]**

Chaldean Version-

[From "Berossus"]

Belus was before **[the creator has always been here]** Belus cut Omoroca in half. **[Omoroca is the "heaven and Earth" super planet]** Half became the Earth and the other half became the heavens

The Zoroaster version

from "Zadspram"

In the beginning Ohrmazd was the light [God] - The sea was scattered which animated the Earth. The Earth was bound together- **[The Earth was the "Earth/ firmament" super planet]** A great rain in the beginning of creation tore the Earth by noise and wind. One portion, as much as ½ the whole Earth remained in the middle and the other portion formed the ocean [firmament] around it. **[The Earth split and asteroids were made.]** These and other examples seem to tell us the same story. The planets have not been in the orbits that are in today and earth was significantly changed by something. It changed from a huge planet to a not quite as large one and it didn't do this "change" in accordance with Bode's Law.

Jewish Version

This comes from the book of Genesis Chapter 1. I'm sure you are confident that the Bible does not tell the same story, but let's read carefully.] God created the "heaven and Earth". **[One way to read this "heaven and Earth" thing is to say it was probably a super planet much bigger than the planet we live on today. According to the progression of the verses, this "super planet thing" was made before God made either the "heaven where angels live" or the "stars/heavens", so it wasn't talking about either "HEAVEN" in this verse. What else could the term heaven in this verse mean?]**

Then the Earth became without form and void. **[The use of the term THEN is interesting. Previously it had form and was not void. This sounds like the "PLANET" had form before it "Split". What else would make it loose its form?]** *Then Darkness [or death] was on the face of the deep.* **[This is a really mixed up verse as there is little doubt that no face "which usually means a turning part" can be found on anything that is dark or deep.**

The word choshek [Darkness] not only means darkness, it also is typically translated destruction, sorrow, DEATH, and obscurity.

The word Deep-It is also very curious that the author used the word deep rather the term for ocean or sea usually interpreted as waters. Therefore the deep is not the sea or ocean.

The word Face-The word "paniym" not only means face. It is also translated to mean- accept, before, against, anger, as long as, at, battle, because of, beseech, countenance, edge, employ, endure, enquire, favor, fear of, for, forefront, from, front, heaviness, himself, in, it, look, me, meet, more than, mouth, of, off, old time, on, open, out of, sight, and upon. Of these, the word face is the least descriptive of what might have happened to the deep.

[A more easily understood translation might be *"Death was seen in the deep."*—I'll go with deep space.]

Then God said let there be light. [The word for light can also be translated as "LIFE". [While this was probably a reference to bringing about life, it certainly was not light from the sun as the writer had not introduced the light from the sun as of yet.]

Then God placed an angel-guarded firmament [barrier] between groups of waters [life givers or planets]. **[Again I have to bring up misinterpreted words.** *Waters is one such word. "Waters" in this verse does not mean water as they were two distinctly different words in Hebrew.* [A probable meaning was "planets". If we look at previous accounts, confirmation of this can also be found in Sumerian description. The firmament, probably the asteroid belt, could separate waters/planets below and above. The asteroid belt was made up of pieces split away from the new planet Earth. It

33

separated and is still a firmament between two groups of planets, Jovian and Terrestrial].

Then God made the dry land to appear in the midst of the water on Earth. ***[In this verse the writer reverts back to the word water instead of waters so it is addressing the ocean and assuring us that "waters" was not "of the ocean".]*** *After that, God made the Sun and moon to shine.* ***[Now we get to the light from the sun.]*** *Finally he* **re-made** *all living things and man.* [The term "RE-MADE" rather than made is significant in this verse and should not be ignored. Animals and men were here before this verse and they had to be made again.]

Modern English Version

Some won't like it but here is a more modern way of saying what is written in the old English version.

God made the Heaven/earth planet which contained life. [Including humans] Massive destruction of life occurred. The destruction affected deep space. After the destruction, God recreated life. [Including humans] He separated one group of planets from another group with a firmament. God remade dry land on the earth. Then God remade all living things and humans.

I know you might not like how I dissected the first part of our Bible, but let me tell you that you have not been really reading that book. You have simply been relying on others to read it for you. I'm not trying to violate an ancient truth, I'm trying to open your eyes so that YOU CAN READ the words and get information for yourselves. That brings me to BODE.

Forget Bode's "Law?"

All the previous histories identified above don't go along with Bode's Law. People have been trying to force fit the planet positions into a neat "Bode's Law" package for years without success. The "LAW*?*" was invented by a man named Johann "Titus" in 1766 so we got the bright idea to call it Bode's Law.

Its name is not close to the inventor's name and it is not even close to being a law.

Some of the planets seem to be close, but when it comes down to it, the positions don't match. They never have and they never will. Our Solar system is still forming. I know it's hard to believe, but even as recently as 12 thousand years ago there appears to have been a major repositioning of the Bode's Law uncontrolled masses. The best thing we can do is to simply ignore it.

Cross it out of your science books.

At the very least, Mars, Earth, Pluto and Venus have all wandered away from their current orbits in very recent times. By building this Bode thing, students have been forced into not even looking at the evidence around them.

Oort Cloud and the Kuiper Belt

Sorry for the outburst about Bode. I will try to stay on track and look at the more distant planetoids. Not to be outdone, the Oort cloud also split around this same time frame. Like the planets, Bode's Law had no effect at all on this split. At least 31 of the planetoids established a somewhat more stable orbit around the sun. The more stable area is now called the Kuiper belt. The largest planetoids in the "Kuiper" group are named Pluto, KX76, Verona, Quaoar, and Chiron. Exactly how they got there is unknown, but interpretations of the Sumerian texts indicate that at least one of them or all of them were moons that were pulled away from one of the Jovian planets as another extremely large planet entered the "known" Solar System from its eccentric orbit in the Oort cloud. I'm not getting into what this planet might have been in this work, but certainly it is possible to have large masses with eccentric orbits in the Oort cloud. These masses could take thousands of years to orbit the sun so we may never even begin to sense a rouge planetoid like that described by the Sumerians. The drawing below shows our new "500 thousand year old Solar System".

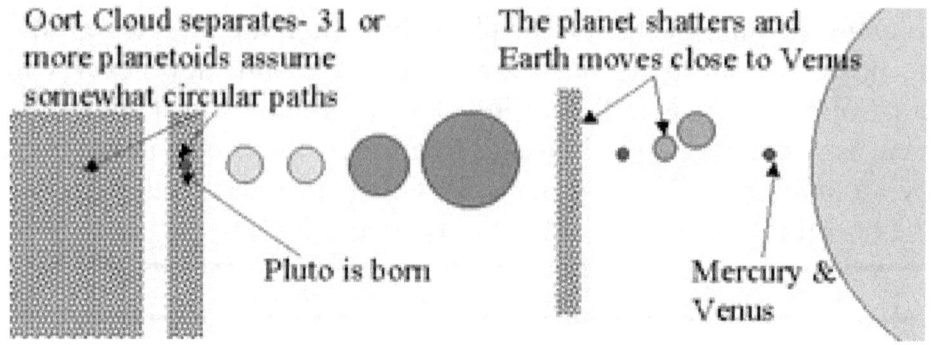

Oort Cloud separates- 31 or more planetoids assume somewhat circular paths

The planet shatters and Earth moves close to Venus

Pluto is born

Mercury & Venus

Solar System Changed Again

While you have been told the Solar System is stable, the next major change occurred only 11 thousand years ago. The evidence tells us that the moon of Venus exploded which sent thousands of particles raining down on the inhabitants of the Earth and set up Venus for a terrible disaster from which it never recovered. I know you have never heard about Venus having a moon and you haven't heard about the disaster that caused it to become superheated happening only 11 thousand years ago. But scientists have discovered the evidence including thousands and thousands of craters that formed on the earth formed; a huge split on the surface of the planet Venus and huge Venusian craters all located on the planet's equator. Everything happened at the same time 12 thousand years ago. A later chapter is coming up that deals in this event in more detail, but don't ignore the evidence simply because no one told you about this in the past. People have been protecting you by building a more comfortable history that has little to do with fact. Another reason to not ignore this key event is that it changed the Earth dramatically. The shifted Solar system is shown below. While it is our current positioning, our Solar System is not as stable as you have been told.

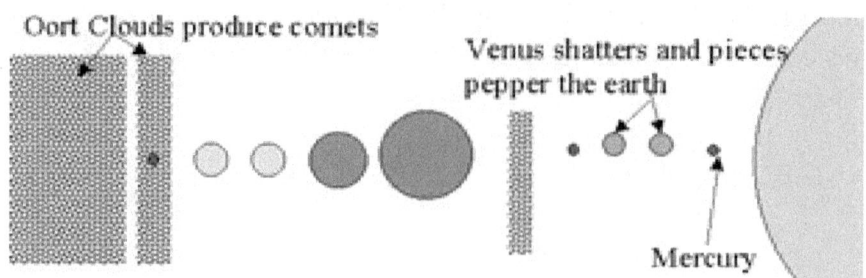

Oort Clouds produce comets

Venus shatters and pieces pepper the earth

Mercury

Since the 11 thousand year ago changes, there has been no significant change in planetary positioning. That does not mean there won't be, it simply means that 11 thousand years is no time at all for a solar system.

Life on the Planets

While I am concentrating on the earth in this book, I briefly introduced some evidence that civilized people lived on the earth hundreds of thousands of years ago. That fact becomes more and more certain every day as more and more evidence keeps pointing to the same thing. Later I will provide more of the evidence because understanding that people lived on the earth during the Jurassic period is a necessity if we are to tie religion and science together successfully. Once one begins to believe that people lived here for a long time it is not too difficult to believe that they would have been able to visit close planets just like we have done today. One should recognize that life was possible on many of the Solar System Planets; especially during the very ancient times. Where there is possibility, one should search for evidence. What we find is a huge amount of supporting evidence. The evidence suggests that life on Mars, Venus, and the Moon was probable and that people living on the earth during ancient times knew about these people living so far away. I'm not getting into what happened to these people on other Planets in this book but scientists are now finding evidence to support the many references---Again let me emphasize that there is a lot of evidence. I will tell you this. If people did live on other solar system planets, those people did not come from beyond the solar system or even from some other planet. The evidence shows that the colonists came from the earth. Many of them never were able to return and died away from their true home.

I know you are getting tired of me saying that "the evidence shows" without showing you the evidence.

Some of it will be readdressed later and some of it cannot be addressed here as it would make the book very tedious. My suggestion is to read my anthology on the development of mankind to get more detail or identify a specific subject of interest and start with the internet for details. Make sure you investigate with an open mind and with the idea that religion and science do go together or you won't get very far. Anyway, let's get back to this line of thinking. Speaking of investigation, you must first find out more about how humans have lived on the planet a little longer than is typically taught in our schools. To make that investigation a little easier, I have included a small sampling of the huge amount of evidence that confirms the existence of civilized humans on the earth many, many thousands of years ago. The probability that humans inhabited the near planets very long ago and left evidence of those "trips" is being increased continuously as new revealing evidence is being uncovered all the time.

Rating Planets

You might think that it is absurd to talk about humans being on other planets, but today, scientists rate planets by water content and atmosphere and earth IS not or was not the only candidate for human life. In fact, there are many planetoids that have adequate components for survival. Below is a list of the "suspected livable planetoids" and there sizes. The planetoids that are highlighted are those **less** likely to have ever contained human life. The others have **or had** an abundance of oxygen and water. More than likely, **people** visited or lived on these planets at one time in our past. When I say in the past I mean many, many thousands of years ago. I know you're thinking, "If people visited planets, then they must have had flying ships and they must have been VERY advanced, hundreds of thousands of years ago." **Both items have been proven to an almost certainty, but that is another story.** *White entries have had all requirements for human life in the past*

Solar System Planetoids by Size

Rank	Name	Diameter [M]	Rank	Name	Diameter [M]
1	Jupiter	71492	9	Titan [S]	2575
2	Saturn	60268	10	Mercury	2438
3	Uranus	25559	11	Callisto [J]	2400
4	Neptune	24764	12	Pluto	2200
5	Earth	6378	13	Io [J]	1815
6	Venus	6052	14	Moon [E]	1738
7	Mars	3398	15	Europa [J]	1569
8	Ganymede [J]	2631	16	Triton [N]	1353

When Were the Planets Inhabited?

I know you don't like simply taking my word for the possibility that people like you and me visited the other planets in our solar system, but trust me on this one and please don't think that little green men ventured into our solar system thousands of times to support all of the VERIFIED sightings of "visitor" ships in the sky. I know it sounds like I'm talking about little green men visiting earth, but I'm not. Instead, I'm talking about ancient humans. In fact, there is little evidence to suggest that humanoids or other creatures live on these planets today. The collected evidence only suggests civilizations were there thousands of years ago. The apparent buildings and artifacts found on these planets could have been left over from just about any time as no major erosion event was on any of the planets besides Venus.

Two things give us indications of age: water and volcanic action. If there were no water now or in the relatively recent past, that would be a good indication that the artifacts were of an extremely distant past. A couple of the hundreds of artifacts discovered on Mars are shown below. Note the relatively low amount of erosion noted. If the planet surface had extreme volcanic disruption, the artifacts would be destroyed very quickly. Here come the scientists to tell us more.

Two Martian Buildings

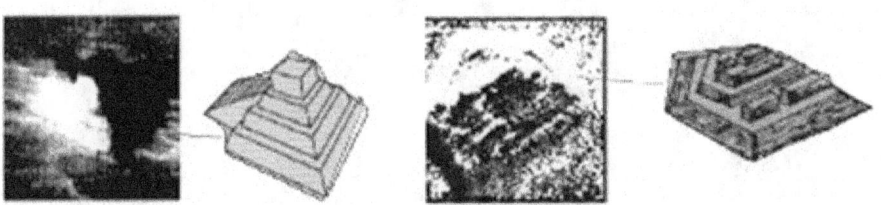

While the pictures are grainy and the resolution is low, I think you can tell that there are regular lines and defined shapes in the structures shown above. Multiply these possible "buildings" by many times and pretty soon it becomes hard not to believe that "someone" once lived on Mars and some of the other planets. The picture below is provided simply because so much has been said about a rock that looks like a face. The picture shows the rock on the right. The probability that it was or is anything is low; however, the group of objects to the left has been studied by many. The regular shapes, just like those shown previously are too geometric to have been cause by natural elements. What you are looking at may very well be the remains of a large city. The area circled is an almost prefect "huge" pyramid left over before Mars became unlivable.

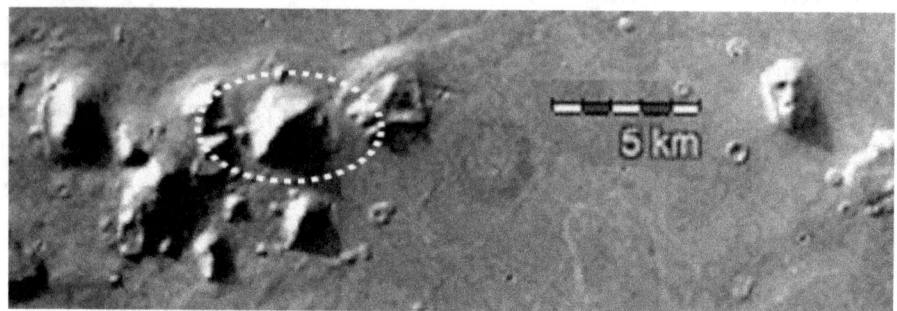

How About the Moon?

If people lived on other planets, wouldn't they have used the moon as a stepping off place??? Sure enough, hundreds of pieces of otherwise anomalous things have been found on the moon. One of the pieces of photographic evidence is of a bunch of shopping centers. Certainly they are not shopping centers, but the buildings are all clustered together on the Moon and show rectangular shapes, Hallways, Rounded domes and all the other things one might expect from a shopping arena. Here is the image below.

If you are interested in seeing more evidence about all the flying to other planets stuff I have collected a good amount of the evidence in another book entitled "The Truth about Flying".

Martian Dirt

The moon is not necessarily a great place to live, but Mars and Venus both were during some portions of their development. Because Mars is the closest planet not having 500 degree temperatures, we know the most about it and may even be able to get an idea of when people lived there. The Martian soil and surface water may tell us when the destruction of the atmosphere and the major portion of the civilization were. It has been determined that some of the dry riverbeds were wet as recently as 20 thousand years ago on Mars. On Venus we have found many dry riverbeds all over its surface. To the left below is one of the probable riverbeds of Mars and to the right is a river on Venus so the answer for Mars and Venus is any time before the water left.

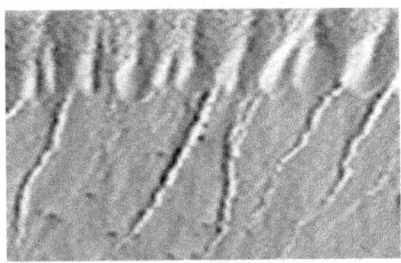 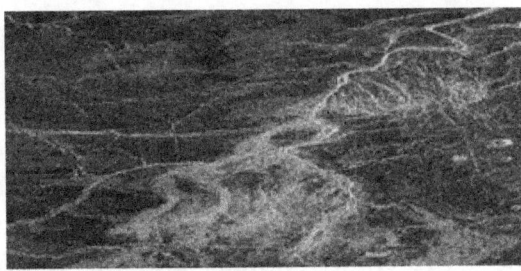

Lunar Water

The Moon could never have been inhabited unless there was water. Sure enough, scientists now know that water is still located at the North Pole. No reasonable dating can be made, but life could have been possible with water.

Mercurian Water

On Mercury they found water on both poles. Here is a weird one. More water is located on Mercury than there is on the Moon. Because its rotation and revolution cycles are so similar, there could have once been life. At the poles there could have been some semblance of life provided that the population slowly migrated to insure that they stayed on the "dark" side of the planet and along the pole.

Just think of it. There may have been life on all of the inner terrestrial planets. This life may have been present as recently as 10 thousand years ago.

Europan Water

Scientists have also found an abundance of water on one of Jupiter's moons, Europa, which makes it also a strong candidate for ancient life. Other planets also have various amounts of water. Don't think that people could not have lived on these other planets because we don't live their today. The planets are changing and not necessarily for the good of man. In the past, living on another planet might have been a reasonable thing to do. Today, the planets are not hospitable and our Solar system is not as well established as you have been led to believe.

Our Solar System Today

Some may not know what our Solar system looks like so I thought it would be good to provide a brief oveview. Most will say that there are 9 planets and earth is the only one that is liveable. Some will say Pluto is the farthest from the sun while others will correctly say that Neptune is farther away from the sun than Pluto. What we don't generally consider are the asteroid belt planetoids, the Oort cloud comets, and the Kuiper Belt planetoids that I just mentioned. Each system is very important in its own way and should not be ignored. The diagram below shows the general conception of what makes up the solar system.

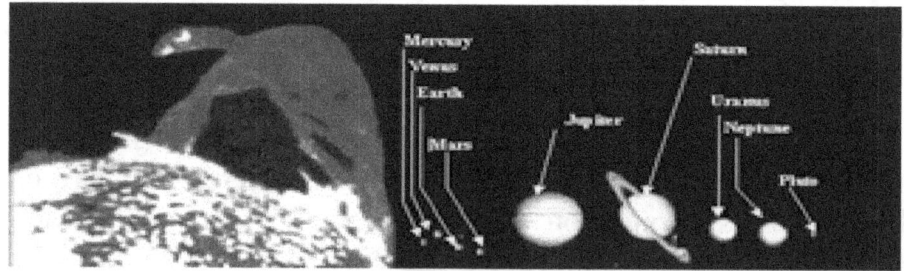

Planets Veruna and KX76

The above list does not make the total sum of major planetoids. Even though you are not typically told about two more beyond Pluto, they have been studied over the last 15 to 20 years and have become our closest known major planetoids beyond the orbit of Pluto. One is named Veruna and the other is simply named KX76. Planetoid KX76 has been tracked now for the past 18 years. It has been determined to be only slightly smaller than Pluto at 1.5 km in diameter and its orbit is very similar in its elliptic shape, but its average orbit distance is 6.5 billion km away from the sun while Pluto orbits at about 6 billion km and it has a diameter of 2.2 km.

Pluto also has a moon named Chiron that is 1.2 km in diameter, so Pluto's moon becomes the 4th largest planetoid in the Kuiper belt of our solar system. Planetoid Veruna is the next largest planetoid with a diameter of about 1 km. I know I missed number 3, but it's coming up. The orbits of these somewhat minor planets are shown below, but if we look even past the Kuiper belt and into the huge pile of dust captured by our sun called the Oort cloud, we may find and even more spectacular find that is typically not brought out. Besides Varuna and KX76, there are many more and additional ones are being found every year. One is named Quaoar.

Planet Quaoar

In June of 2002 the Pluto "Is it a Planet or not controversy" got more real as the new planet Quaoar was discovered and this new spherical mass is a much better planetary example than Pluto. Although it is only about half the diameter of Pluto, its orbit that identifies this orbiter. Unlike the more comet-like orbit of Pluto and its doublet Charon, along with the orbits of Veruna and KX76, Quaoar's orbit is similar to that of the other planets and almost circular as shown below. By adding Quaoar, Veruna, KX76, to the "normal" 9 makes at least 12 major planets within the Kuiper Belt or closer to the sun. [I know the Asteroid Ceres with a diameter of 650 miles is also a giant orbiting mass almost as large as Veruna, so it could also be added if you want to.]

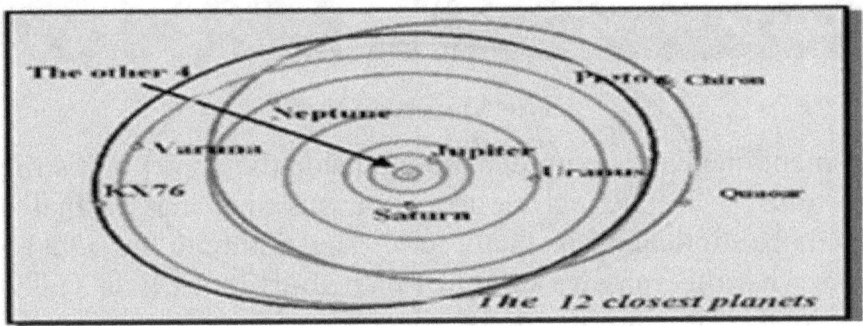

If you want to stick with only nine, there is good reason to throw Pluto out and replace it with Quaoar as our ninth planet. Quaoar is roughly spherical and about 1,280 km in diameter, about one-third the size of our moon. It is greater in volume than all 50,000 of the numbered objects in the asteroid belt combined, minus Ceres. Ceres is a special case, which by itself comprises 1/3 of the total mass of the asteroids in the asteroid belt.

Quaoar is probably comprised of equal parts rock and ice and has temperatures of 230 degrees Celsius below zero. The ice is a rock-hard cocktail of frozen compounds or gases including methane, methanol, carbon dioxide, and carbon monoxide. It has all the building blocks, but none of the heat required for life. One Quaoar year is 288 Earth years so even a one year old should be respected for both his stamina and his age. Its circular path around the Sun is six billion kilometers away from Earth. That is 1.6 billion kilometers further out than the mean orbit of the Pluto-Charon pair, however, much of the time Pluto is actually farther away from the sun than Quaoar due to its elliptical romp around the solar system. Pluto, KX76, Ceres, and Veruna possibly aren't true planets, but Quaoar should be considered our ninth member. Next is a size comparison chart. The distances shown are the various diameters of these often ignored planets.

 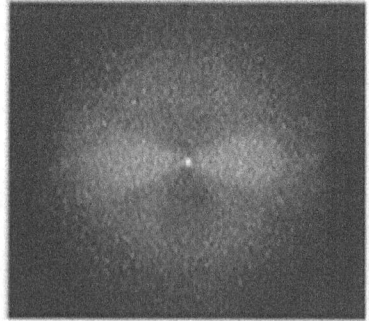

The Massive Oort

As I mentioned above, immediately outside the kupier belt string of planetoids is a spherical cloud of various size masses that have even more eliptical orbits than Pluto. The picture to the right above shows what this massive group of planetoids looks like. This Oort thing makes up almost all of the space identified as our solar system. The bright area in the center would be our "Normal" solar system and the huge "haze" arount it is Oort.

Many of these planetoids are called Comets as they show off a tale as their orbits come closer to the sun. Some believe that very large masses in this area which extends out 50 thousand times as far from the sun as our Earth, about 5 trillion miles, could contain some form of life. Our children should not be confined to think of the Solar System as a group of 9 planets, especially not with Pluto as the last one. The other thing children should be presented with is the probability that Mars not only had people living there at one time, it also molded our earth.

Mars Formed the Earth

Mars helped form our planet more than plate tectonics. Plate tectonics is a real and fundamental component of our earth's development, but smashing "Plates" did not make the huge mountain ranges like you have been told. There simply is not enough energy supplied. Plate tectonic theory is grossly flawed when it comes to the major mountain ranges of the world and the earth splitting phenomenon identified above cannot explain the long lines of mountains either.

Besides making mountains, Mars also made our moon and the Pacific Ocean, but first let's examine our mountains.

Uplifted Mountains

Just think about it. The mountain range that makes up the western shores of the South American and North American Continents and goes around to the other hemisphere making some of Japan and other pulled up areas was, by plate tectonic <u>misconception</u>. Supposedly they were made from massive crustal plates smashing into one another. Let me just ask this one question. If the mountain range goes ¾ the way around the earth, "WHAT DIRECTION WERE THE PLATES GOING?" Aha! You can't tell me because it is impossible. I did some other calculations below to further address this completely unsupported "theory" or at least what some people call a theory.

This "Uplift", well tested theory, on the other hand, provides the <u>strong evidence</u> that Mars pulled up the major mountain ranges of the earth.

Let me give you a brief synopsis of this probable occurrence. As indicated previously, the solar system components were not always in the orbits we have today. Earth and Mars were much closer at one time and interacted with one another on occasion. When I say interacted, I'm not kidding. As the two planets came close to one another on one occasion, the gravitational pulls from both planets caused mountains to appear along the equatorial line of both planets. That sounds like a lot of gravity, but the mathematics has been tested and there is no doubt that it is both possible and probable if the orbits had been much closer than they are today, and yes, there is very good reason to believe that the planets were very close neighbors. The biggest reason is the lack of craters on half of Mars, and that will be explained very shortly.

One such "pull" caused the pulled up area that makes up the western coast of North and South America and continued around the other side and "yes" the earth axis was not where it is today and Antarctic was positioned at the equator. A second time mountains were pulled up along the middle of China and India. This time the two planets had been even closer and the mountains were pulled even higher. Mars had made almost all of the major mountain ranges on our earth and it had gotten pretty wrinkled itself.

Plate Tectonic Anomaly

I know you are still trying to believe the plate tectonic thing and are fully convinced that this work is not a true characterization of what happened because someone would have taught you differently if there was truth in my words. For a minute let's go back to school. From the time you were in grammar school all the way through college you were continuously told that "plate tectonics" was the action that formed the mountains of the Himalayas, reasonably called the Himalayan Ridge. This same plate tectonic thing caused the long range of mountains called the American Ridge that cover the western side of South and North America. We were told so many times, that we didn't question its absurdity. Below is a graphic of the idea. The plates simply went past each other and all the dirt and gravel that was on top built the mountains

If the Himalayan ridge is the intersection of two "Plates" and one plate was rammed against another huge mass about 200 million years ago such that "**more than a quadrillion tons of Earth**" were

pushed up an average **"2 miles into the sky"**, it took more than a few little volcanoes to push the slab with the force required. It would have taken a cataclysmic event we can't even imagine. Then, supposedly, the same thing happened again for the American mountain ridge.

Estimates have been made using volcanoes as the initial tectonic pushing agent that **the two "volcanoes" required to do the "mountain pushing" would each have been <u>about the size of the United States</u>**. Just imagine the complete absurdity. There just isn't any evidence of these huge explosions. The whole concept is so ridiculous now, that I'm embarrassed I ever believed the tale in the first place. The problem is that the science community started with fact. Plate tectonics has been proven and evidence of its existence is seen every day. Then they went berserk because they couldn't figure out what caused these huge piles of rock. Instead of letting our children know that there are **"unknowns that should be explored"**, we continuously insulate, stagnate, and destroy the minds of our children. We sort of liquefy their thought process so that it can be molded into thinking the way some group of our society wishes people would think.

Wrong Equator

You may have recognized that a mountain range being pulled up in a straight line could only have occurred along an equatorial path of a planet. Certainly the path along the edge of the Americas could not be the equator. Certainly the line demarking the Himalayan ridge could not have been the equator. The whole concept must be false and yet the mountains are there just the same. Later, I will show you that there is STRONG evidence that or equator is continuously moving around and when it moves, it is a catastrophe for the earth and everything that lives on it. There is absolutely no doubt about it but our teachers don't tell any of our children because it is a VERY uncomfortable fact.

Not a Conspiracist Statement

With all these rouge ideas about how educators are teaching our children lies you might think that I am a Conspiracist. I am not and I don't even think the educators know that they are doing this dastardly deed, but that is one of the main reasons I am writing this book—to get people thinking for a change. Hopefully the information will also get to our vulnerable children.

Pull Up Review

Thanks to many satellite pictures and computer models, we now know that the Mountain ranges actually formed along the equatorial perimeter during some cataclysmic event and the mountains were pulled up into place rather than having two dirt piles pushed together as multitudes of volcanoes generated pushing power to move gigantic plates around and smash together like a demolition derby. Something terribly bad happened to Earth and Mars and there is a high probability that they happened at or near the same time. By the way, this is not some hair-brained concept that I pulled out of my head. Many reputable scientists are currently doing a lot of research in this area. These scientists also are investigation a third and final encounter.

That Second Encounter

That second encounter of Mars and Earth was a disaster. Mars and Earth again came into close proximity and began pulling up the Himalayas but as the earth rotated around a little more both exploded as indicated previously. We can even be pretty sure about the time of this occurrence as <u>1/2 Million years ago</u>.

Here is the outcome of this last and worst encounter. The earth had a huge chunk pulled out where the Pacific Ocean is today and Mars lost almost half of its surface. Five of the largest craters on earth have been dated to that same time and a massive extinction has been recorded in the history of earth. Additionally, the Atlantic expansion and separation of Pangea has been dated to that same time and the crustal regeneration at the bottom of the Pacific has been dated to about 1/2 Million years ago using the newer dating methods identified earlier. Earth survived the catastrophe, but just barely. Mars on the other hand did not fare well at all.

Mars Should Have Split

Even with all the evidence of the 1/2 million year old catastrophe many still say—bah Humbug—to this whole concept. The near collision of Mars and Earth must be false because physical law requires that the smaller planet would have sustained the greatest damage. As I have already explained, THEY ARE RIGHT; at least in their theory. Let's look at the remains of Mars for a minute. One thing to notice about the Mars we see today is that only half of the planet has significant amounts of cratering. Mars definitely got the worst of the encounter. It essentially split in half. That fact is so obvious, it is almost comical.

Thin Crust on Mars

The land under the Pacific Ocean on earth has almost no crustal mass remaining after the split, but Mars is worse. Over the ENTIRE northern hemisphere of Mars, the crust is rarely more than a few kilometers thick and the sparsely cratered surface is strongly suggestive of a relatively new surface. Like the remaining portion of the earth, not including the Pacific Ocean, the southern hemisphere of Mars has a strikingly thick crust, which exceeds 20 kilometers in places, and a much more heavily cratered surface. It is in this hemisphere that we find nearly all the major impact basins such as Hellas, Isidis and Argyre with crater basins well over 1 thousand Kilometers in diameter. These huge holes were probably made by some of the large chunks of earth that left during the explosion. As the graphic shows, after its last "near collision" with earth, Mars also became a new planet, much smaller than it had once been.

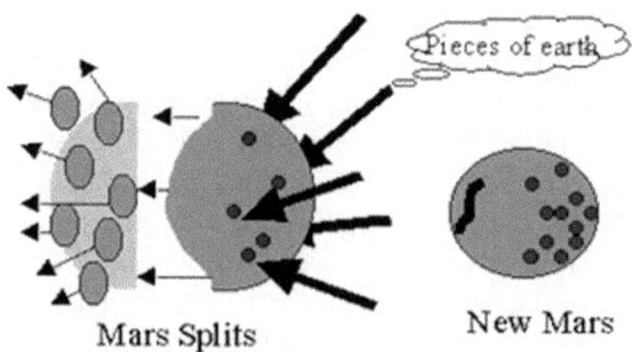

Mars Splits   New Mars

Evidence of a Split Planet

This odd cratering isn't most obvious evidence. Here is the proof from NASA. I know all of this sounds bizarre, so let's look at a topographical image of Mars. Please note that the northern hemisphere is not only smooth, but it also is sunken in much worse than our Pacific Ocean. It has a mean surface height <u>6 thousand meters lower</u> than the mean of the southern hemisphere. Where do you suppose the northern half of the Planet went to?

Half of the planet being smooth and sunken and asteroid debris nearby might get you to wondering, but shouldn't there be some "tear" areas that didn't completely let loose as the planet was splitting? The answer is YES.

Split Mark

Also look at the dividing line between the hemispheres. Just like the huge slashes under the Pacific known as Mariana's trench, the northern hemisphere of Mars, is marked by one huge gash called the Valles Marineris. Although it is only 7 kilometers deep, it is up to 200 kilometers wide in places and has a total length of 4,500 kilometers. It is almost like Mars was split apart at one time and one of the marks of the split is this huge gash just like the Mariana's Trench in the Pacific, but many have not recognized the obvious.

Healing Mars

Mars, like the earth, is slowly trying to heal itself from this ancient event. In another 200 thousand years, the entire surface will probably be about the same height and here on earth, the Pacific Ocean will be about the size of the Atlantic. The diagram below shows the general topographical distinction on Mars at the present time. The darker area is the high area and it is slowly resurfacing the planet. As it does, a hole is opening at the South Pole.

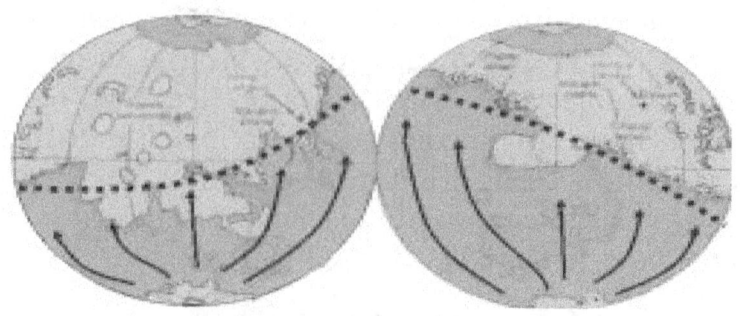

Pangea Was Not the Only Continent

The "Pangea Partner Theory" provides the answer to the anomalous question of how a single super-continent Pangea could have formed in the first place. The answer, of course, is it could not. **<u>Surely you were told that in school!!!!</u>** Earth mass must be equally distributed. It's sort of a law of equidistant continents or something like that. Anyway, the earth will tend to establish that condition, no matter what. The only way for Pangea to have been here is that another super-continent was located where the Pacific Ocean is today. There simply is NO other possibility. There is no telling how many animals were yanked away from the earth when the Pacific Ocean was made.

Pangea Anomaly

We have all been taught about how Pangea contained all the land mass of the earth and slowly separated 450 thousand years ago, but there should be one nagging thing in your head.

Why was all the landmass of the earth clumped together in the form of Pangea in the first place?

With all the land mass in one tiny, tiny area, the roundness of the earth would have been questionable, but the size of the earth dictates that it will establish a spherical shape. The spherical shape would have had to have been established millions of years before Pangea was formed. If we believe that Pangea existed; and there is great amounts of data to confirm its existence; we are left with an unanswered anomaly.

Before the big chunk of earth left, there **must** have been land generally around the earth to support the roundness requirement and the basic truth of Pangea's existence.

I'll say again, the whole Pangea thing makes no sense unless there was an equally large mass on the other side of the world. Where did it go? I think you know! You should be getting pretty angry about now at the lies that have been told you? I know the historians and scientists have all lied to make our history more pleasant, but that isn't what is needed.

Dating the Creation of the Pacific?

As I stated, the Mars incident occurred only about 1/2 million years ago as the Triassic Period began. To test that answer we need to look at dinosaurs. The period from the end of the Triassic until the end of the Cretaceous is known for one thing—dinosaurs—BIG ones. At the end of this time, almost overnight, things got much smaller. There is a reasonable probability that one of the initiators for the large dinosaurs was the creation of the Pacific Ocean. Everything was great for the dinosaurs until something happened. The evidence tells us that the end of the age of dinosaurs was quickly climaxed by the slowing down of the earth's rotation. I know those claims are not presented in other works, so let me start by listing things we think we know. I'll get back to the light animals later, so don't worry.

The Earth Changed in the Triassic Period

If nothing else, the Triassic boundary is distinctive. Something happened to put a hole where the Pacific Ocean is today and this same event caused the super-continent of Pangea to begin to pull itself apart, and killed most of the living organisms on the planet.

According to data collected from the Deep Sea Drilling Project [1968 to 1983] the Pacific Ocean Basin was determined to be youthful, possibly less than a million years old. This was determined by testing mantle depth and sedimentation over the ocean floor compared to the average mantle depth.

According to some models, the break-up of Pangea occurred around 1/2 million years ago. One simple test can be accomplished to test this date. The Atlantic Ocean is the major separating line of the super continent and it is still getting wider by about 3 centimeters per year with a current average width of about 5,000 kilometers. As the Atlantic is already about 1/3 the width of the Pacific, we can believe much of the pressure to fill in the Ocean has been relieved and an average spreading speed of about 1 meter per year would be an reasonable approximation Therefore, the close encounter and the explosion that made the Pacific Ocean would have occurred around 1/2 million years ago.

Asteroids Again

Let me revisit the asteroids for a minute. I indicated that the asteroids were probably formed by the close encounter of Mars and the earth. Some may say how they could be so very far away from the earth if that was the case. Situated between Jupiter and Mars, the asteroids don't appear to have been the debris from the earth encounter, but another blasted planet. Here are some problems. If they don't represent the encounter, where is the earth's missing Pacific Ocean land?

The Blast material had much less dense than the earth or Mars as it did not contain the Core materials. Because the density was lower and the momentum came from the planets themselves the revolutionary spin would have slowed drastically and with it the orbital positioning would have moved well beyond the Mars orbit. Think of it like an ice skater spinning like a top. While his density was lower and he has arms flailing, he spins slowly. Pulling in his arms and legs to increase his density makes him spin much faster. The asteroids would have reacted opposite to that action. Their sudden drop in density would have flung them far away.

Let me just say; the date of the Pacific bottom, the date of the Pangea split, and the date of the most extensive extinctions together point to the most traumatic event of our earth's history--the last encounter with Mars. This encounter did some bad things, but it may have been the reason for the development of huge dinosaurs, because it may have made things lighter.

Everything Used to Be Lighter

Have you ever wonder why the animals of the Jurassic, Triassic, and Cretaceous periods were so big? I'm sure you were told that evolution caused the gigantic monsters and then evolution caused them to die off. Well I'm sorry to disappoint you and say that that is rubbish. If evolution had been responsible for the change, then the change would have continued. There is a much simpler answer and for that answer we should first look at ice skaters again.

Try to picture our ice skater spinning on an ice rink; if he wants to go faster, he simply makes himself smaller. He pull's himself into a tight space and he spins faster and faster. As he makes himself wider, he slows down. The earth did that very same thing 1/2 Million years ago.

When the earth lost the continent over the Pacific Ocean, there is little doubt that it began to spin faster. When I say little doubt, I mean the obvious. Animals got bigger. Even the insects got larger during this "special time". If the earth began to spin faster for a time, everything would become lighter and that is what makes things get bigger for no other "apparent" reason. Don't laugh at this theory, the evidence is pretty solid.

Dinosaur Anomaly

If you have wondered why flying reptiles "supposedly" were not able to fly and why the dinosaurs "supposedly" could not run because their heart could not pump blood fast enough, the scientists reporting these findings did not help sooth your concern. The answer that they should have gravitated towards is that the flying dinosaurs **were** able to fly and dinosaurs were able to **run**. They were simply lighter than they would be if they lived today. If you are wondering why the oxygen levels are not as high today as they were 1/2 million years ago, the answer again is a faster spinning earth. Don't let some "artificial scientist" tell you that a flying dinosaur had to crawl up the side of a mountain so that it could open its wings and glide down to the ground hoping to snare some food on the way. I guarantee you that the species would not have survived. Don't let that same type of "scientist" tell you that a dinosaur could not run after his food. That animal would not survive. Don't let ANYONE tell you that a diplodocus could not lift his long extended head. If a dinosaur could not lift his head, he would be dead.

Fast Earth Theory

After saying that things were lighter, one must try to understand how this very strange effect occurred so there is an important corollary I call the Fast Earth Theory. I know that the statement seems fanciful, but you can see its reasonableness if you look at Saturn, Neptune, and Uranus today. All of them are spinning so fast that things on their surface are much lighter than they should be. Even the atmosphere is lighter and all three are losing that valuable commodity known as atmosphere every day. This will continue until they slow down just as the earth did. This important theory provides a plausible reason why huge land dinosaurs don't exist today, but were able to exist in the past.

Everything Got Heavier

The Martian event may have been a catalyst. When the earth got smaller and more dense, it began to spin faster and everything got lighter, but it did not sustain that spin rate.   It is apparent that about 1/2 million years after the near collision, the earth began slowing down its rotation rather over a short period of time. I know that's another wild idea that hasn't been brought out by others so let me provide you with the evidence.

Dinosaur Size Evidence

During the Jurassic, Triassic, and Cretaceous periods, the largest animals that ever walked on land were in abundance. The bone structures of these massive monsters don't exactly fit the animal structures that would be required today. If there is an anomaly it probably means that we haven't uncovered the truth.

One such anomaly is shown with the Diplodocus. Its neck is too long. If the creature had tried to stretch out its neck in front of its body, the head would have come crashing to the ground as there is not enough muscle to pull it into the air unless its neck was lighter.

The Tyrannosaurus Rex structure is such that it would not have been able to run in our atmosphere. I know some of the new theories have the Tyrannosaurus Rex as a scavenger, but that would only work if he could run. If it couldn't run it would have died.

Many of the Bird-like reptiles would have surely died as they could not fly in our current atmosphere. To make this element more comical, scientists have invented many theories that suggest that the winged reptiles would have had to climb up to the top of a cliff and take off into the air. If they ever hit the ground they would be doomed as they could not regain flight. I know it sounds stupid, but that is the belief of many today. Hopefully you can appreciate the absurdity of such a theory.

Still another problem is the dinosaur heart. It has been determined, by many, that the hearts on these massive beasts would not have supported the blood flows needed to sustain body function. Many have surmised that the dinosaur's blood flow problem could have been rectified by increased oxygen percentage in the air. The blood flow could be greatly reduced if more oxygen was present in each breath, but that does not account for many of the physical characteristic anomalies, and I'm not sure it would even satisfy the blood flow problem without other elements like lighter animals.

Scientists have proven that the oxygen content was higher during the time of the dinosaurs by simply sampling the oxygen captured in globs of amber. Therefore you have the question of why the oxygen was here 1/4 million years ago and not here now. Think I can bring some sanity into this whole crazy mess.

The answers are that animals were lighter in the old days because the earth quickly lost some of its size. It also had more oxygen in the atmosphere at the same time and lost much of it because it was spinning too fast to hold onto the oxygen for the same reason.

Less Density Evidence

If we make the assumption that the specific gravity was less during these ancient times, then there must have been a cause and we don't have to search long. Generally speaking, less gravity means smaller size, less density, or faster rotation. I don't believe that the earth got larger at the end of the Cretaceous period, so the other two seem more likely. Concerning density, we have one very nice piece of evidence—the Moon. Below are three of the elements that we believe we know concerning the Moon and its relationship to the earth.

The Earth has a large iron core, but the moon does not. The Moon density is only about 50% of the earth's. This is because Earth's iron had already drained into the core by the time the moon was yanked away.

The moon has exactly the same oxygen isotope composition as the Earth, whereas Mars rocks and meteorites from other parts of the solar system have different oxygen isotope compositions. This shows that the moon, most likely, formed from material in Earth's neighborhood.

The moon orbit and Earth rotation are synchronized, suggesting that they both came from a common source.

I have presented an argument that the moon came from the earth, so let's just say that the Pacific Ocean opened up and spewed out the moon. The moon density level tells us that only the lighter portions of the earth were expelled, leaving the much heavier mantle intact. Therefore the specific gravity associated with the earth was much higher after the expulsion than before it and Dinosaurs would have been lighter before the Moon was made than equal sized animals from after this extreme event that reduced the earth's density.

Because today, the Moon is only 50% as dense and the Earth, we can make a reasonable guess that the density of the earth was perhaps 10 to 15 percent less than it is today before that section was torn away. Therefore the dinosaurs would weigh as much as 10 to 15% less than they would today because of this factor alone, but that would not be enough difference. There must be another factor to be considered.

Faster Rotation

The lower density wasn't the only thing that lightened the dinosaurs. We also have a pretty good example of what the earth might have been like with respect to rotation if we look at a picture of Saturn, shown next.

Flattened Planet

Saturn is wide at its equator because it rotates quickly. The Earth may have looked this way before the end of the Cretaceous. Saturn is spinning fast enough so that particles do not have to gain substantial amounts of energy to attain what is sometimes called escape velocity. Saturn's day is less than ½ of a "current" earth day and the whole planet is trying to escape. The bulge at the equator shows two things. It shows that components of the atmosphere are drifting away from the planet and it shows that gravity is lower. Remember the "atmosphere drifting away concept for the topic below". What we find in our solar system is that, even though Saturn and Uranus are substantially larger than the earth, things on those planets are lighter than they are on earth. This is due to the high speed spin.

On Saturn or Venus, animals would weigh only 90% of what they weigh on the earth and on Uranus they would only weigh 80% of their "Earth weight".

One reason we can believe the earth was spinning fasted is the thicker air that has been lost over time.

Thicker, Oxygenated Air

We can also surmise, from evidence described below that before the earth slowed down, it probably had a thicker atmosphere rich in oxygen. The most logical way for the oxygen level to have slowly been reduced over time would be that our planet rotation was significantly faster than its present rate and similar to that of Saturn. Here is what researchers have found concerning the oxygen levels during the dinosaur days.

Experiments have been done to show that the oxygen content of the "Dinosaur's Air" was higher than our current air. This was easily accomplished by simply testing the oxygen content of captured "air" in ancient resins. The tests showed a substantial increase in the oxygen content when compared with "Modern Air". The problem with this experimental data is that the researchers only used the data to try to prove that a "water canopy" was around the earth before Noah's worldwide flood. Because the canopy theory had other problems, the issue was essentially disregarded.

Because the air was thicker, the earth was most likely spinning faster to allow the oxygen to leave and other elements fall into place.

Flying Reptiles that Flew

Remember the flying reptiles that couldn't fly? It would even be easier to fly and the flying reptiles wouldn't die if they touched the ground. Imagine that! Flying reptiles actually flew. Also the Archaeopteryx must have flown and its wings were not useless as you have been told.

Why Would the Planet have Slowed Down?

There is a simple answer to this question. A huge meteor hit the earth and saved our planet, but the Dinosaurs were destroyed. They weren't destroyed by the K-T boundary that everyone talks about and the volcanic action or even the reduction in sunlight and reduction in plant life as a food source. They were destroyed by their own weight. I know there were heavy animals after the meteor strike of at the end of the Cretaceous, but generally their bulk had to be supported by water and the land creatures after that fateful time were limited to the size of an elephant.

Even without Mars messing things up, the earth had its own internal problems. Its iron core was not and is not stable. Therefore, the rotational axis can "slip" from time to time. As it slips, hot spots form and explosions abound. Added to this is the continuous bombardment of meteors and comets along with the occasional exploding moon and you have an unstable earth.

The Earth Goes Wild

The earth has been molded by many things besides Mars, so we should look at some of them to get a feeling about what the earth will do today and in our future. Some of the elements of effect are volcanoes, plate shifts, the axis of rotation flipping as shown below and meteors. Each is described below.

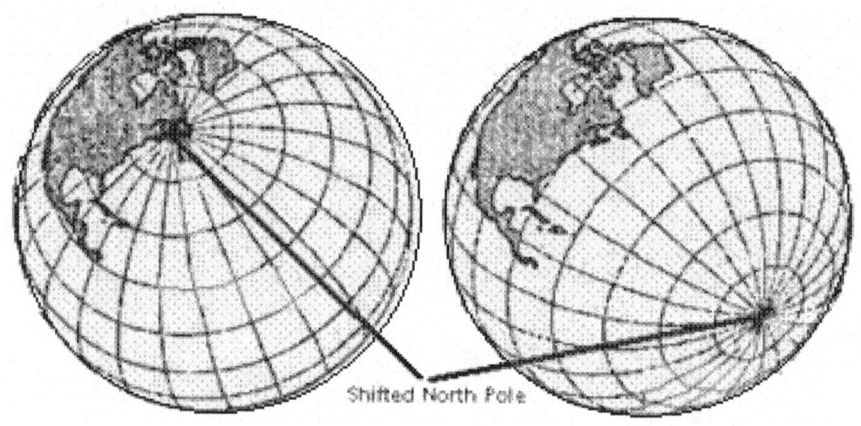

Shifted North Pole

Volcanoes

While there is no evidence of the volcanic action needed to force mountains into the sky, that doesn't mean there weren't some big ones. Researchers have determined that there are peaks of volcanic action in earth's history that have occurred throughout time. At the end of the Pleistocene, end of the Jurassic, end of the Triassic, and end of the Permian Ages mark some of the worst volcanic periods recorded. From time to time the earth just blows off steam. Maybe tomorrow will be that day. Who knows. Below is a 2004 map of the world and the major volcanoes are shown as white dots. Our Earth is full of volcanic holes. Each one is ready to spread what is inside the Earth onto the outside. Some of the holes are spewing lava at this very moment. They will never make a major mountain range, but they can be very deadly.

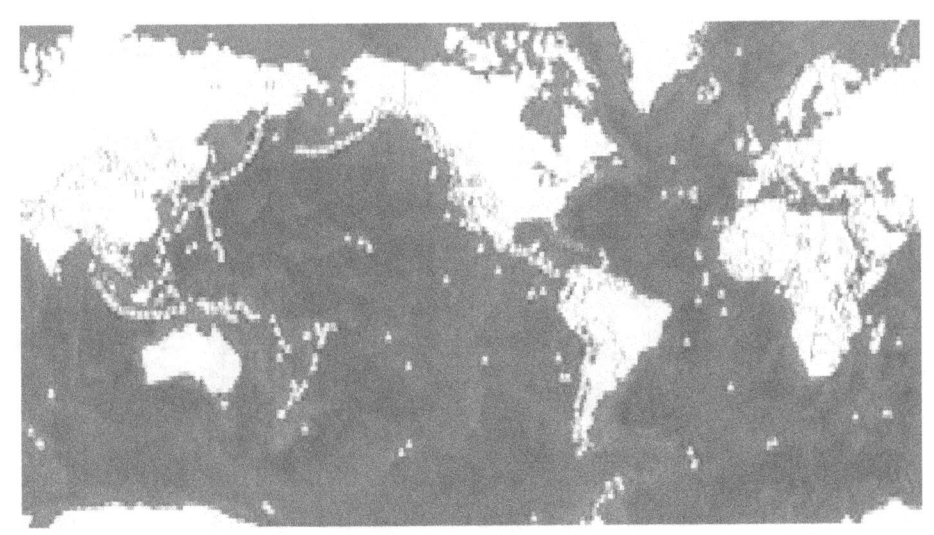

Polar Axis Flips

This well tested theory provides striking evidence to support the "earth flipping on its rotational axis" phenomenon and its devastating effects. Haven't you wondered why they found thousands of animals quick frozen in Alaska and Siberia? The answer has nothing to do with some instantaneous Ice Age. It has to do with a warmer area quickly becoming the North Pole. Here's the scary part. According to Atlantic Ocean sea bed magnetics testing and piles of other evidence, this whole flipping thing happens a lot. One reason for Ice Ages is the erratic nature of the earth itself. You may not want to hear this, but the polar axis flips about every 100 thousand years or so and things get destroyed when it occurs. The graph following shows the last 14 flips over a period of 400 thousand years using Potassium-Argon nuclear decay dating of magnetic material in solidified magma in the center of the Atlantic Ocean. Another flip could happen any time. Using mathematical models of the external crust and inner molten material, researchers have estimated with mathematic models that the Earth should flip on its axis about every 100 thousand years. The problem with trying to determine the actual workings of the Earth is that no one has ever seen the inside of the Earth to model it properly, but the results do confirm the high possibility of a polar flip, which will cause mass destruction, tidal waves, and major climatic changes. A polar flip, however, does not cause the most damage. Crust movement or magnetic field wander causes the real bad problems.

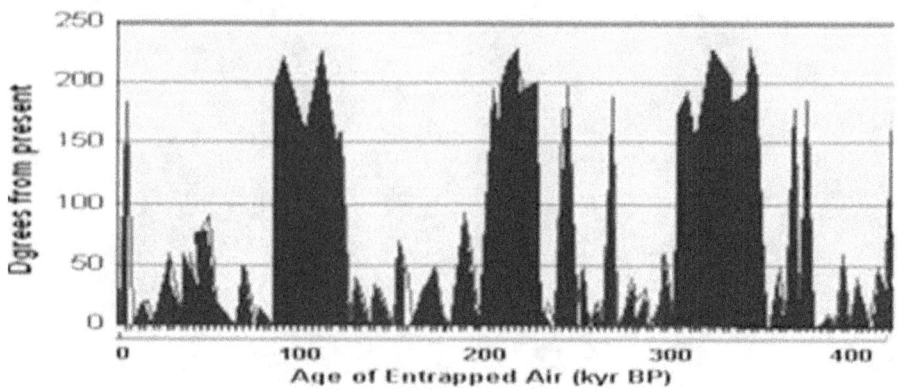

Note that in addition to the 180 degree flips of the earth's magnetic field, there are many exertions that did not result in an axis flip, but instead, the earth temporarily assumed an oblique rotation. This is called Magnetic Field Wander.

The last major movement occurred about **9 to 12 thousand years ago**. This did not, by itself, immediately affect most things, but did cause increased winds and a magnetic pull different than that, which would have been felt prior to the flip. This change in magnetic field together with the Earth crust movement, at the same time, could have made the jump even faster than normal which caused even more catastrophic destruction. That time stamp is an important one so remember it for a while.

"Scientific American"-Magnetic field reversed more than 170 times in past <u>80 million years</u> [using nuclear decay dating] last reversal, according to Nature and "New Scientist" and "Scientific American" magazines, occurred 13 thousand years ago. **The next shift is expected 2030AD**.

Atlantic Ocean-Cooled magma in the center of the Atlantic ocean caused by the Atlantic ocean getting wider every year shows that the magnetic field of the poles have shifted and wandered many times in the past.

In the graph preceding, each jerk represents substantial stress on animal life. Whenever the movement is shown to move by 180 degrees, the magnetic field actually flipped.

Plate Shifts

Another way to look at magnetic wander is called plate shifts. Like the magnetic shifts, these apparent crust movements have been estimated to happen about every 20,000 years. One researcher indicated that the most recent ones occurred 43,000, 22,000, and **10,000 years ago**. Sometimes the crust and magnetic field seems to wander over a number of years and other times it seems to jerk suddenly. In fact, the evidence suggests that there have been at least 170 major movements in the Earth's crust, which corresponds to the magnetic field shifts during the recoverable paleo-magnetic timeline from the Atlantic Ocean. One of the theories is that these "jerks in the crust are apparently caused by the uneven weight of the various plates supported on the surface of the Earth; especially the 19 quadrillion tons of mass called Antarctica which is located at the present day South Pole. Each time a movement occurs, terrible things happen like tropical areas turning into glaciers. Whether the evidence shows magnetic field wander or plate shift wander doesn't really matter, because the outcome is the same.

Tropical Arctic

Researchers have found evidence that the Arctic was tropical during the Jurassic Era. As they investigated the depths, they found bones of early crocodiles, turtles and fish that were all tropical and estimated the summer temperatures reached into the 90s. The most logical explanation for the hot temperatures was that the plates shifted or the planet axis moved by a substantial amount many years ago. Finds similar to this have convinced many that the outer core of the Earth moves continually and that the movement is in jerks over time.

Tropical Antarctic

If we move to the other side of the world, we find the same thing. The remains of tropical trees were found among the Magma beds from some ancient time. Additionally, they have found swamp type dinosaur bones along with remains of swamp type plants that existed before Antarctica became cold.

Hot Spot Proof

While this was the lead overview of the book, here it is again for more detail and reinforcement. If the axis is changing, there should be some dramatic physical evidence and there is. The evidence is not only from the magnetic field alignments of molten material in the Atlantic Ocean, but also some easily seen evidence. The evidence is in the form of hot spots. The best hot spot to discuss is Hawaii. The volcanic action in Hawaii has nothing to do with the edges of the plates. The picture below shows the basic outlines of the major plates and these anomalous "Hot Spots". The hot spots don't stay still. They wander, but they wander in straight lines interrupted by abrupt turns. By measuring the distance the "hot spot" travels, we can determine how long the Earth or a particular plate on the Earth stayed with a particular axis of rotation. The hot spots wander because the inner core is much more dense that the outer core, and occasionally the two slip in the direction perpendicular to the axis of rotation. The reason we know the slippage is perpendicular is that it is still happening.

Plate Movement Direction

If we look at the apparent trail of the Hawaiian Islands over time as shown below. Under the ocean a clear path is noted and times for each abrupt change has been approximated by distance. By the way, a new hotspot has just opened 73km south of the big island showing that the plate wander direct is still in the same direction as it has been over the last 10 thousand years and as I stated above the direction is perpendicular to the Earth rotational axis.

I guess you noticed that I indicated that the last shift in the axis happened only 10 thousand years ago. How and why that all happened is a story in itself, but the story will have to wait a little while longer.

The Hawaiian Island chain isn't the only hot spot group that shows this pattern. Look below and see that two other hotspot wander directions in the Pacific Ocean look similar to that of the Hawaiian Island spot.

Some Don't Believe in Shifting Poles

Some people try to infer that this whole thing about the Earth changing its axis is hogwash. A twisting earth that could completely change in a moment's notice and kill millions of humans and animals could not possibly be a correct model. Unfortunately, there is just way too much data to assume otherwise. Antarctica with its dinosaur bones, the quick frozen Mammoths, the various polarities of the deposited iron from volcanic action in the middle of the Atlantic Ocean; they all tell the same story. The Earth axis can move and with it there can be relatively fast and devastating climatic changes. These changes are horrible, but may not be the responsible party for most of the extinction periods. The most effective exterminator on the Earth has been and will continue to be the Comet or Meteor. Whenever a comet or a major meteorite storm hits and the earth axis of rotation shifts right afterwards, total chaos occurs as it did 10 thousand years ago.

Thousands of Meteors

Many of the extinction periods were brought on by elements outside our planet. One element is called a meteor. Many thousands of these meteors have hit our planet and scientists have built up a pretty good picture of major meteor strikes just by dating the craters they leave behind. The graph below shows 14 known major meteor hits, which caused unbelievably huge craters that are greater than 50 kilometers [30 miles] across. Imagine what would have made such a hole and imagine the destruction. There is more than a coincident timetable between extinction periods and the time of these meteor impacts and I don't mean just the one that we have all heard about that struck the Yucatan and signaled the end of the dinosaur age. I mean almost every destruction period was preceded by a huge meteor attack.

The graph shows the immense size of some of these major impact craters. Some are over 300 kilometers [180 miles] in diameter. Whenever the meteors hit, a thin film of iridium dust sometimes would cover the surface of much of the Earth and allow scientists to categorize and time the event very accurately. Iridium is not found in abundance on the Earth except at these layers, so it's a really nice timing device.

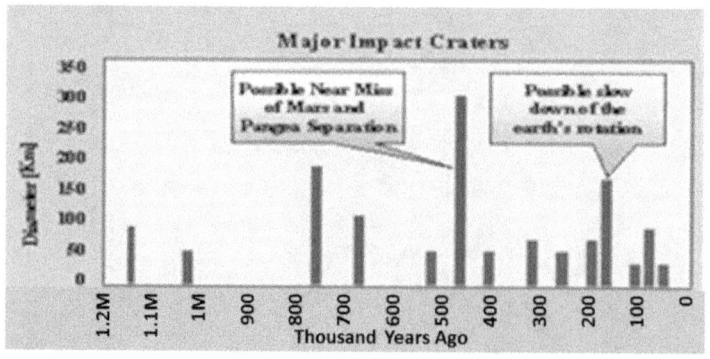

Below are 28 of the largest "known" impact craters found to date. Knowing that crater impacts on the water are much harder to find, you can image how many HUGE meteors have hit the earth.

Major Cratering Around the World

When [T]	Period	Where	Crater Dia. [KM]
60M	Archeo.	South Africa	300
60M	Archeo.	Ontario, Canada	250
4M	Protero.	Antarctica	250
3M	Protero.	Montana, U.S.A.	60
1M	Cambrian	Australia	90
950	Ordovic	Sweden	52
850	Silurian	USA	100
750	Devonian	Australia, Chad	40 & 125
650	Carbon.	Quebec, Canada	54
650	Carbon.	Canada	35 & 25
550	Permian	Antarctica	50 & 40
450	Triassic	Canada, France	100 & 250
350	Triassic	USA	50
250	Triassic	Russia	80
200	Jurassic	S. Africa, Norway	70/ 75
130	Jurassic	Australia	55
120	Cretac.	Mexico & Russia	170 & 45
60	Tertiary	Nova Scotia	45
35	Pleisto.	Virginia, Siberia	90 & 100
25	Pleisto.	Tajikistan	52

Each of these huge meteors, not only had enough force to cause local extinction and massive cratering, but they also many had enough impact force to split open the Earth. Of course, I don't mean that the Earth was split in half many times, but there is strong evidence to indicate that the meteors caused great openings in the crust which allowed huge piles of magma to be expelled. That discussion is covered later, but just image a meteor that hits so hard that the earth begins to split open.

450 Thousand Years Ago End of Triassic

Let's look at the meteors that hit 450 Thousand years ago in particular. So far we have categorized at least 5 major craters that occurred from a massive planetoid that split up just before impact on the Earth during that time and "yes" the planetoid evidently was Mars as we looked at earlier.

The impact craters can be found in France, two in Canada, the Ukraine, and North Dakota. Talk about destruction! That was probably the worst destruction ever. Over 90 percent of all creatures became extinct and that was just one of the times that an immense, impact crater, maker apparently caused mass extinction as we examine coincident timeframes.

Meteor Evidence

The previous graph shows of the major impact craters over time along with their relative sizes. You can probably already begin to see a correlation to the previous destruction table and this crater graph. One of the craters not shown on the graph occurred about 3 million years ago and is the largest of all **known** craters on Earth. [See below left] It is located in Antarctica and is 1/2 mile deep and 150 miles in diameter. This event probably caused some significant troubles on Earth as this estimated 13 billion ton meteor stuck at a speed of 44 thousand miles per hour. The event was so very long ago there weren't any known animals to witness an extinction so let's concentrate on extinction meteors.

The Chad Crater, [left] which is over 100 kilometers across. The Meteorite storm that caused it also caused a major extinction 750 Thousand years ago. Two chains of craters have actually been found in the area showing that whatever hit shattered into many pieces and struck many sites. This was the largest piece.

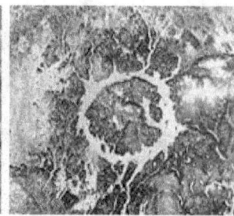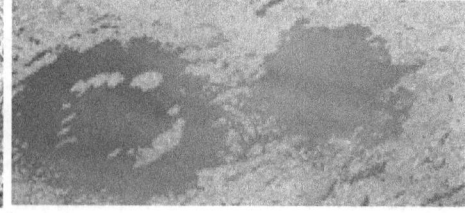

The Canadian crater shown second is over 100 kilometers across and is known to have been one of many that occurred about 450 Thousand years ago. It was part of the major group of meteorites that hit about the same time, as indicated previously.

Another Canadian crater set-The picture [previous right] shows two of the craters formed over 450 Thousand years ago one is 35km and the other is 25km in diameter. These could be remnants of the last great flyby of the planet Mars.

The infamous Yucatan crater to the right occurred whenever the dinosaurs were destroyed 120 Thousand years ago. It is over 170 kilometers across. This particular "hit" caused one of the iridium layers we use for dating and this layer extends around the entire Earth. The Yucatan meteor gave us a lot of iridium and took away those pesky dinosaurs. The only thing remaining is a perfectly round indention half on land and half in the Gulf of Mexico.

The Yucatan crater is special in that large animals ceased to exist at the time it hit around 120 Thousand years ago. As discussed earlier, the earth may have slowed its rotation and halted the loss of oxygen from the air at this time. Everyone got heavier and the loss of oxygen into the Solar System ceased. This level of stabilization spelled disaster for any huge animals. [See below left]

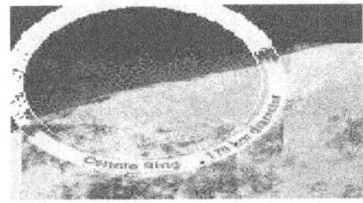 

Antarctic Bone Bed

The impact of this giant meteor some 120 Thousand years ago left a calling card besides Iridium. Fossil deposits on Seymour Island, Antarctica tell some of the details. A giant bed of fish bones at least 50 square kilometers in area can be found there as if some sort of catastrophe had annihilated untold millions of fish. And you guessed it! This great bone bed was deposited directly on top of that layer of extraterrestrial iridium that marks the 120 Thousand - year-old event. The much, much smaller [only 10 miles across] and well defined **Barringer Crater** in Arizona occurred in the fairly recent past [40 thousand years ago] and probably did little major damage, but it is still impressive. [See above right]

Where do Meteors Come From?

One of the most dangerous impact crater making strikes come from Comets. As the comets heat, large pieces of debris break away and form sort of a cloud around the huge mass of icy material that typifies a comet. The particles orbit the sun with the comet and when a comet comes near, watch out. Below are some of the more deadly comets. They are deadly in that they come closer to the earth than others. Sometimes comets even hit the Earth. As you can see, many of the comets come extremely close to the sun. Some get closer than the Earth is to the sun. Wirtanen and Schwassmann will be coming extremely close in the next few years. The year 2013 is of particular interest in that the Mayan calendar was abruptly halted at the end of 2012. Whether they knew something about this event is not known, but why they would have picked that date is surely a strange one.

Close Encounters

Comet common name	Revolution [Years]	Most recent or next occurrence	Closest location in AUs	Magnitude comparison
Halley	76.1	1986	0.6	5.4
Encke	3.3	2003	1.3	9.3
d'Arrest	6.5	2005	1.3	8.3
Tempel 1	5.5	2005	1.5	12.3
Borrelly	6.9	2001	1.4	11.4
Giacobini-Zinner	6.5	1998	1	9.3
Grigg-Skjellerup	5.1	1992	1	12.3
Crommelin	27.9	1984	0.7	12.3
Honda-Mrkos	5.3	1995	0.5	13.3
Wirtanen	5.5	2013	1	9.3
Tempel-Tuttle	32.9	1998	1	9.3
Schwassmann	5.4	2006	0.9	11.3
Kohoutek	6.4	1973	1.6	12.3
West-Kohoutek	6.5	2000	1.6	10.4
Wild 2	6.4	2003	1.6	6.4
Chiron	50.7	1996	8.5	
Wilson-Harrington	4.3	2001	1	9.3
Hale-Bopp	4000	1997	0.9	1.3
Hyakutake	40000	1996	0.2	

Sometimes the Earth Splits

The shifting earth causes many problems, but so has the occurrence of some of the largest pieces of meteorite material. The meteors sometimes hit so hard, the earth splits open.

This is not a theory, by the way. It's a fact

Many places around the world are nothing more than huge piles of magma that have surfaced after the earth split open. The Deccan area of India, for instance, is made up of 12 thousand cubic MILES of LAVA piled up to make mountains. There is no doubt about it. In the past, large meteors have hit so hard that the earth split open. This catastrophe has happened quite a few times and the evidence is astounding. Millions of cubic MILES of magma have spewed up in mountainous piles at the sites of the splits. The magma mountains can be seen around the world and, guess what!; scientists have known about them for a long time, but no one teaches the scary facts.

Critics say that meteors alone would not cause the kinds of extinctions that have been witnessed over time and they are absolutely right. In addition to huge meteor craters, many times the Earth was actually split open by meter strikes or similar action. I know that sounds scary, but evidence is evidence. These splits typically caused **millions of cubic kilometers** of lava and debris to spew out. The smoke and debris from this action sometimes covered the Earth with material that could block out the sun. What I think is the most enlightening part is that the split usually occurred on the opposite side of the Earth from the meteor hit. Just think about how much force would be required to do that. Here are some of the major lava flow events that have occurred since about the Jurassic Period. The events I'm talking about here are not the occasional volcanic eruption. These are the effects of the Earth splitting apart in a huge area and spewing out **millions** of cubic kilometers of lava. The graph shows some of the major events that caused lava flows in excess of 500 thousand cubic kilometers.

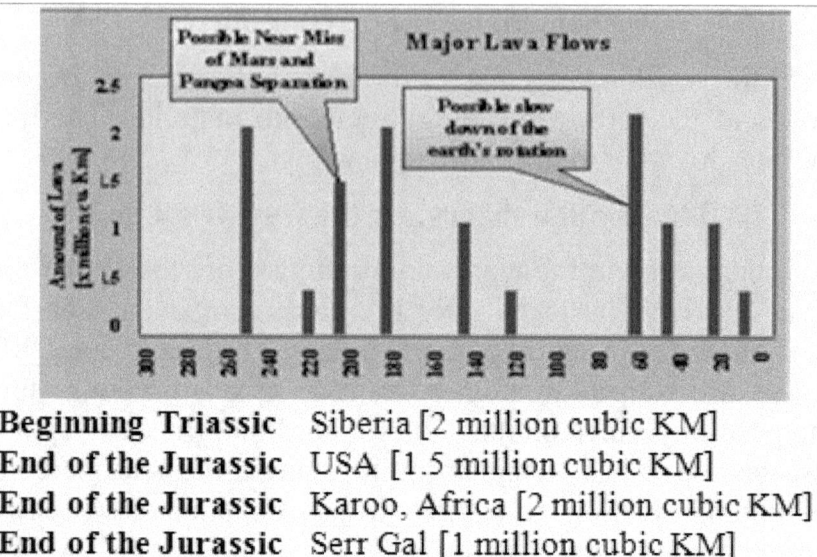

Beginning Triassic	Siberia [2 million cubic KM]
End of the Jurassic	USA [1.5 million cubic KM]
End of the Jurassic	Karoo, Africa [2 million cubic KM]
End of the Jurassic	Serr Gal [1 million cubic KM]
End of Cretaceous	Deccan, India [2 million cubic KM]
End of the Tertiary	Ethiopia [1 million cubic KM]
Pleistocene Age	Ross Sea, Antarctica [1 M cu. KM]

The picture below shows one of the largest piles of lava in the world. It is in India, in fact it used to cover just about the whole country and it still covers over 200,000 square miles and is 6,500 feet thick as shown to the right. It is estimated that originally it covered 1.5 million square miles and contained over 12,000 cubic MILES of lava. [**That's right I said MILES, not meters**] When the Earth splits open it splits wide open.

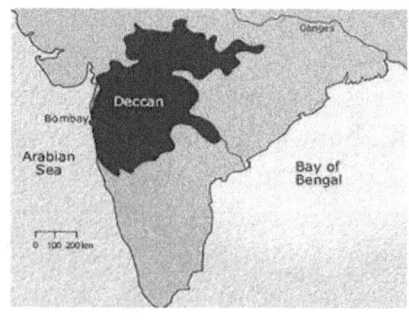

As I mentioned before, one of the times that it split open, a huge chunk actually split away from the Earth to make the moon. The Pacific Ocean was also created at the same time. By the way, even when these huge lava eruptions occurred, no new mountain ranges were formed except by the piles of lava. And if that force wasn't enough to push two of the plates fast enough to cause the Himalayas just what was supposed to have caused the great plate smashes? Schoolbooks typically don't cover these huge lava flows because someone might ask that very question. They also don't talk about the Earth splitting open because someone might get uncomfortable when thinking that the Earth could possibly come apart. Unfortunately, it could and being comfortable doesn't change reality, my friend.

Are Meteors Harmful?

Some believe that, except for very rare occasions, meteors don't really hurt anything. What about this list below? This is just a small sampling of the many disasters caused by meteorites in recent history. These things fall all the time and many times people are killed.

Meteoric Damage		
When	**Where**	**Reported effect from Meteors**
616AD	China	10 deaths from meteor shower
~1350	China	Iron rain killed people & animals
1490	Shansi, China	Stones fell like rain, 10,000 dead
1511	Lombardy, Italy	Monk and several animals killed
1639	China	Large stone fell, 10s killed
~1640	Indian Ocean	Two sailors killed
~1650	Milano, Italy	Monk killed
10/30/1801	Suffolf, England	Home set on fire and destroyed
1/16/1825	Oriang, India	Man killed; woman injured
12/11/1836	Macao. Brazil	3 homes damaged, oxen killed
5/1/1860	Concord, Oh.	Horse killed
6/30/1874	Chin-kuei China	Child killed, house crushed
1/31/1879	Dun, France	Farmer killed
3/11/1897	Martinsville, W.V.	Horse killed and man injured
6/28/1911	Nakhla, Egypt	Dog killed
9/5/1907	Hsin-p'ai, China	Whole family crushed to death
6/30/1908	Tunguska, Siberia	2 killed, many injured
7/19/1912	Holbrook, Arizona	14000 meteors, building struck
12/8/1929	Yugoslavia	One killed in a bridal party
11/28/1954	Alabama	Woman hit by 4kg meteor
8/14/1992	Mbole, Uganda	48 stones fell, one boy hit

Can It Happen Again?

From the thousands of planetoids of significant size that are being tracked, beyond those classified as comets, we can make another short list of the ones we expect to come close to us in the near future. Chances are, something will hit us in the near future. It might be 10 years or it might be a thousand years, but we will feel the assault of these intruders.

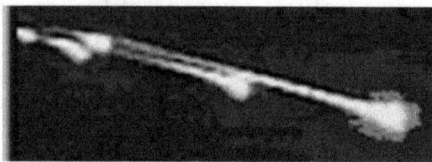 

If one splits into five pieces like this one over Peekskill New York or the second one seen over Texas, we can just be thankful as the smaller pieces pepper the ground. Following is a list of those which will pass within 900 thousand miles of our planet, unfortunately the accuracy of the tracking is only about +/- 900 thousand miles until the objects get relatively close to us, so who knows. Note also that some come close to us often. The Planetoid named WN5 is one such nemesis which has a "near" collision scheduled in 2039.

Collision Asteroids			
Name	When	Closeness [AU]	/[MILES]
AD2	2133 Apr.	0.0087	800,000
AN10	2027 Aug.	0.0027	250,000
CU11	2080 Aug.	0.0043	400,000
DB7	2048 Feb.	0.0079	700,000
DU3	2143 Mar.	0.0067	600,000
DV9	2160 Jan.	0.0066	600,000
Hathor	2069 Oct.	0.0066	600,000
Hathor	2086 Oct.	0.0060	500,000
HH49	2023 Oct.	0.0079	700,000
LV	2076 Aug.	0.0071	600,000
MN	2010 June	0.0076	700,000
Nereus	2060 Feb.	0.0080	750,000
Nereus	2166 Feb.	0.0057	500,000
NN4	2144 June	0.0048	450,000
NY40	2002 Aug.	0.0035	300,000
OX4	2148 Jan.	0.0020	180,000
QK130	2128 Mar.	0.0097	900,000
RQ36	2060 Sept.	0.0055	500,000
TU28	2121 Apr.	0.0065	600,000
TU28	2102 Apr.	0.0079	700,000
UG11	2008 Nov.	0.0091	800,000
VP11	2086 Oct.	0.0060	550,000
WN5	2039 June	0.0015	125,000
WN5	2028 June	0.0044	400,000
WO107	2093 Nov.	0.0079	700,000
WO107	2140 Dec.	0.0005	50,000
XF11	2028 Oct.	0.0062	550,000
YB5	2002 Jan.	0.0056	500,000

The possibility that a huge comet or meteor will strike us in the very near future is compounded by visions provided to us from the famous 15th century "seers" Mother Shipton, and Nostradamus. These modern seers and others have projected the outcome of a very near term disaster. I will discuss the probable future comet or Meteor strike later. Before that terrible time, some precursor meteors have fallen.

Our Last Major Meteorite Strike

No one seems to talk about this event, and it happened in a very remote section of the world, but it could have been very noticeable if it hit anywhere else in the world. The date was **December 9, 1997.** At 5:11 A.M., crews of three trawlers at widely separated sites off south Greenland reported "a blazing fireball that turned night into day." At a distance of over 60 miles away, the flash was compared to that from an atmospheric nuclear explosion. Seismic tremors also emanated from Greenland, so the impact of a large meteorite is almost certain. So far, no one has found the remains of the huge meteorite, but you have to recognize how very desolate and impossible that area is to search. Luckily our last major meteor hit Greenland and not Disneyland or people would easily accept the event as a real danger. The 1997 event was small in comparison to those that will happen in the near future. One of the most likely, near term, candidates is about 2 Kilometers wide and it is classified as NT7.

Eminent Asteroid Strike

One asteroid not on the list is one named NT7 was just discovered in 2002 and is on an impact course with Earth. If you thought the others were scary, this one is expected to strike or come extremely close to our planet on 1 February, 2019 and even though it is only 2 kilometers wide, it will be traveling at a rate of 28 thousand miles per hour and contains enough energy to cause continent-wide devastation on Earth, so we should not ignore this terrible threat. To make things worse, if it doesn't hit then, this rock circles the sun every 837 days, so it will have another shot later, but it's not the only chunk of rock that is probably heading for us. To make things look really bad scientists now track a huge quantity of these asteroid masses. The pictures below show those that are close to the earth. Each dot represents a tracked asteroid. The first shows the big picture out to Jupiter while the second one shows tracks items just out to Mars. The circles are the orbits of the major planets close to the sun and a portion of the orbit of Mars is shown as the outside circle. What I really wanted to show here is that the probability of getting hit is higher than most want to believe.

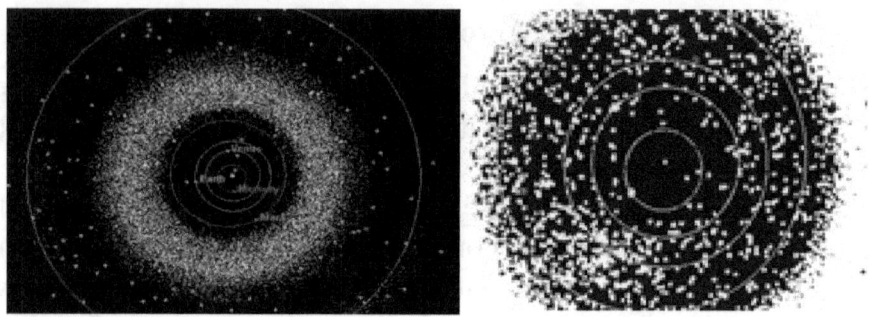

Besides tracked data, we can use two things that are a little more exotic to find out about upcoming meteoric disasters. The first thing is to know that history repeats itself. Many meteors and even comets have hit in the past and they will hit again. Although tried and true, that method has no timing accuracy or reasoning. The second method is by listening to prophets or seers, but most who claim to foretell the future do not. Some special people seemed to have been able to foretell the future and we will need to investigate further.

If one cannot see the future, one should, at least , be able to understand the past. One thing to understand is that while all of the catastrophes on earth were happening, animals flourished. The expansion of the animal kingdom was way too fast and furious. Some say animals got here by some random "survival of the fittest " mutation redone millions of times over millions of years, some say God created everything in 6 days, some say the animals were seeded by aliens from another world. I say let's look at the evidence. First we will investigate the misinterpreted Creationist Theory of Evolution by God's direct creation in 6 days a little more.

Creationist Theory Problems

Before I get into this one, let me first say that I strongly believe that the Bible is written in an accurate and sequentially significant way, but its time base is event driven or relation driven rather than absolute time driven. Some try to force fit the absolute time into secular dogma and when there are problems, very interesting devices are used to reestablish the "Absolute timeline". One of those attempts is now known as the "Creationist Theory".

The Creationist Theory is not really a theory, but is instead, a hypothesis "derived" from a single book of the Bible. It has many problems that are pushed under the rug to keep it going. In order to keep the belief that the world is only 6 thousand years old as this group would determine, one must use an extremely limited interpretation of the Biblical texts. We will see that the evidence does not support this type of theory; nor does the Bible for that matter. Here are some of the elements that show that the time base of this theory is way too short.

Magnetic Field Change Evidence

As I mentioned before, many magnetic field changes have been verified by lava samples from the Atlantic Ocean and other sites. As the planet plates separate, lava pours out in abundance in the middle of the ocean, and the metallic portions of the lava align to the current magnetic field of the Earth before it hardens. Core samples from the bottom of the Atlantic show that there have been at least 170 major changes in the magnetic field alignment but none have occurred in the last 2 thousand years. A process was developed which uses these metal alignments for dating purposes. It is called Paleo-magnetic dating and it does not confirm a short time span. Each time that the magnetic field changes, havoc and destruction is rendered on the inhabitants of the Earth. By creationist standards, the 170 shifts would have occurred in 4 thousand years of improbable and unbelievable massive wobbles, destructions, and annihilations and, during all this mess, the Biblical story has almost no mention of the upheavals except for one that occurred 7 days before the worldwide flood. We will discuss in detail the destruction periods that are revealed in the Biblical texts, but they do not account for what is found in the Atlantic.

Worldwide Flood Evidence

The worldwide flood could not have created the thick pockets of coal, as implied by the creationists. Even if the flood produced wide and uncontrolled manufacture of coal deposits, the amount and depth of the deposits around the world could not possibly have been generated during that one major Earth trauma. In some areas, the coal deposits are **well over a mile thick** and are over an extremely wide area. The thought that the trees all congregated in one spot as they floated around in the flood and then collected to form these massive coal deposits is not probable. Trees must have been in an area, then died, grew, died again, grew, etc.---for thousands and thousands of years.

Physical Similarity Evidence

The physical characteristics between the shorelines of South America and Africa, along with many other indications, strongly suggests that there was once one major continental mass that has been separating slowly over many, many years and it could not have happened over the 2000 years of creation before the flood without showing a major difference in the crustal density in the middle of the Atlantic Ocean. The crustal density is identical in the Atlantic and on the continental masses. Only under the Pacific Ocean is there any major variation in density of the crust, which brings us to another problem.

Thin Crust Evidence

The Earth's crust is thin under the Pacific Ocean. By using seismic mapping techniques it has been determined that the thickness of the crust at the bottom of the Pacific Ocean is much, much less thick in comparison with the thickness of the crust around the rest of the world. As I mentioned previously, this strongly suggests that the Earth was split apart in ancient times and is slowly healing itself. By examining the amount of crustal matter that is deposited each year, the age during which the Earth was split apart has been estimated to be over 450 thousand years ago. If God simply made the Earth and the moon six thousand years ago, why would there be the timing anomaly? I think I know what you would say. The dating methods are no good.

Oxygenated Air Evidence

It has been suggested that the air was more oxygenated long ago and therefore, the decay process was modified before the flood, more oxygen should mean more carbon-based items. This would mean that there would be a higher concentration of carbon 14 than we currently see, which would in-turn mean that any carbon-14 dating that crossed the high oxygen boundary would be in error--- This error would indicate that things that were actually much older would test to be less old because there would be too much carbon-14 remaining. Items tested to be 6,000 years old may be twice as old, which of course is the wrong direction for the creationist belief.

Too Many Animals Evidence

There have been so many animals on the Earth that we still haven't run out of the oil that was produced by the decay of their bodies. If the animals were only here during the 2 thousand years before the flood, the Earth must have somehow been much, much, much larger to allow them to walk and not be piled on top of one another. I know that some believe that oil came from some other means, but right now let's just assume that it came from dinosaurs, because it is the most logical without additional insight. For the billions of barrels of oil to have been deposited, the animal life lived here a long, long time.

Karoo Evidence

This is one of my favorites. On the southern portion of Africa lies the greatest find of terrestrial vertebrate fossils [mostly swamp dwelling reptiles]. It is estimated that there are **800 Billion;** yes that's billion with a B, animals in a sandstone and shale deposit that is **20,000 feet thick.** It is stretched out for hundreds of miles. This simply could not have been a single clump of creatures pushed into one area as the floodwaters subsided. Just imaging a pile of 800 billion animals all clumped in one position after the worldwide flood rotting in the sun. Pee!! You!!

Differences versus Time Evidence

There are too many differences and not enough time. According to Creationist view, the flood occurred 4000 years ago and only olive skinned Adamic people survived. Within a period of about 100 years, they mutated into red skin people, white skin, yellow skin, black skin, brown skin, straight hair, curly hair, flat nose, high cheek bone, and slanted eyed variants around the world. Then, for the next 3,900 years nothing happened at all. Carvings from thousands of years ago and today show people look the same.

Evolution Problems

Those first 100 years must have been something if we are to believe a flood date of 4 thousand years ago and only Noah descendants as the survivors. While the quasi-religious hypothesis is full of holes, the "normal" science hypothesis is even worse. If we looked up the definition of "inappropriate testing and determination" or "inappropriate science", we would find this thing called evolution and survival of the fittest. It is a neat way to show existence of animal life that causes no fear or thought by the evaluator. The problem is that, most of the time; it doesn't work and doesn't fit the evidence. At best this unproven, untested determination should be called a hunch, but it is taught in schools as almost a "Law". Let's look at some of the problems with evolution. Many call evolution the "No-God Creation".

Too Many Animal Types Evidence

Evolution REQUIRES a long drawn out back and forth of mutation to make new creations. While that concept didn't fit into ANY of the evolutionary timelines, someone came up with a harebrained concept which allows for hundreds of useable mutations to occur within a few years and then NOTHING would happen for 10 thousand years or so. All this was to happen "randomly".

Is that ridiculous or what? If we are talking about randomness, then stick to it. If it doesn't work---ABANDON the concept and find a new hypothesis. Of course no one has and this "PACKETIZED evolution" is being taught to our CHILDREN.

Sorry for the burst of emotion, but we need some sensitivity and we should make sense even if we don't like the outcome. Some might say that let's just get rid of the stupid packetizing thing. The problem is you can't. Too many animal types were spontaneously generated if a real evolution theory was to be supported. We will see that after each destruction period, a huge influx of animals was apparently created within an extremely short period rather than the supposed "start over" that evolution theories would require. Each of the seven generally known major destructions recorded in history caused the extermination of well over 80% of all species of life and each time more animals emerged. Here is the strange part. The evolutionary life cycle didn't start over each time, but instead, many creatures **spontaneously reappeared**. To make things even stranger, you will see, in the next chapter, that a complete extinction record shows as many as 18 major extinction periods and many periods of minor extinctions during Earth's life cycle, which further exacerbates the problem.

Quick regeneration of widely diverse life after extinction doesn't go along with evolution.

Out of Place Objects Evidence

For the evolution concept, many things are out of place in time. Man's footprints with dinosaurs, cups found in coal, walls found deep in the Earth, and many other artifacts clearly show that humans were on the Earth many years before the time that evolution can support. A sampling of these anomalies can be seen in later chapters. Don't be fooled into thinking that the artifacts are proof that the world is very young and testing methods are a sham. We will investigate some of the methods and reasons why a young Earth is not a better answer than the "evolution theory".

Out of time relics disprove evolution.

Missing Link Evidence

No crossover animals have been found. No matter how much searching, no one has found any evidence to suggest that a change in genus kind is possible- a horse has always been a horse and a man has always been a man. I know that many are saying that they have found plenty of crossover species like the partial bird partial reptile creatures [Archaeopteryx], but they truly do not represent a cross over. They really describe different separation of species that will be discussed later. If we were looking for crossover people we might look at Charles Darwin himself as referenced to the Homo Habilis drawing to the right. Although they are similar, it does not mean Darwin is a missing link.

Natural Selection Dilemma

Natural Selection and survival-of-the-fittest models don't work. Shell fossils show that during some ages they get big and then small and then big again at a later time period. Horse "evolution" also shows the same characteristic. The horse started off small, got medium sized, got smaller, and then larger. Some may remember seeing the evolution of the horse showing the tiny eohippus and continuing to the modern horse as a gradual size increase over time, but the graphic was a lie. The graphic below shows the historical lineage of the modern horse and how its evolution doesn't make sense. All other animal fossils suggest that the adaptation to the environment does not increase the capability of an organism to survive, nor does it make a superior organism, it just kills the organism off and miraculously a new type takes its place. Sometimes it's a better animal, but <u>many times it is a "less survivable" creature.</u>

Types of Horses and When They Were Here			
Yrs ago	**Type**	**Size**	**Toes**
55M	Hyracotherium	12in.	Four
50M	Orohippus	14in.	Four
47M	Epihippus	20in.	Four
40M	Mesohippus	24in	Three
35M	Miohippus	30in.	Three
24M	Archeohippus	18in.	Three
17M	Merychippus	40in.	Three
16M	Protohippus	28in.	Three
15M	Hyracotherium	40in	Three
14M	Pliohippus	44in.	One
12M	Dinohippus	39in.	One
10M	Astrohippus	48in.	One
4M	Equus [modern]	62in.	One

Long Nose Dilemma Evidence

Sometimes "evolved features" are recognizable as mistakes. A Dinosaur mistake to be considered would be the long nosed Dinosaur. As shown in the picture his nose is over four times as long as his head and was curled back on itself. It was completely useless and there is no evidence to suggest that other dino-features evolved from this mistake. The nose couldn't be used as a battering ram like the horn extensions of other animals. It was just a long nose. For those who would suggest that this is the father of the elephant with his useful nose/hand, it would be improbable that the thing would have evolved from this characteristic to the dangly one of today. Here is where I get into trouble. This type of mistake probably didn't come from an omnipotent creator, nor did it come from an evolutionary process. We will talk about a third option that actually makes sense.

PARASAUROLOPHUS
The Long Nosed Dinosaur

Evolution Reversal Dilemma Evidence

Why are no fishes now changing into amphibians, amphibians into reptiles, reptiles into birds and mammals, and monkeys into man? If growth, development, evolution, were the rule, there would be no lower order of animals. All animals have had sufficient time to develop into the highest orders. Many have remained the same; some have deteriorated, none have evolved.

If plants and animals all developed from a one-celled animal, such as the amoeba, why did the amoeba not develop?

Animal Variety Dilemma Evidence

We are told that, excluding the insect and microscopic world, there are about 3,000,000 species of plants and animals today. About 1/3 of that number are animals and about 0.05% of the animals are going extinct every thousand years. That gives us about 1 million species of animals with over 500 species being lost every thousand years. Now let's assume that the last major extinction took place about 10 thousand years ago. From that comes a very serious question.

How many new species should have arisen in the last 10,000 years to support the undirected evolution theory?

If we start from 10 thousand years ago and assume Noah carried 50 thousand animals on his little boat and he was the only one, there would be 8 animal species doublings to approach the 1 million animal species we have today. If we assume that each doubling takes the same time period, then the last 500 thousand animal species would have sprung into life within the last 1 thousand years not including animals that become extinct so the number would be more like 600 thousand over the last 1000 years. The number is very close to zero.

Sounds like we are getting ready for one of those packetized bombshells where everything mutates into something strange again..

- Darwin indicated the following: "In spite of all the efforts of trained observers, not one change of species into another is on record."

- The Canadian Geologist, Sir William Dawson said: "No case is certainly known in human experience where any species of animal or plant has been so changed as to assume all the characteristics of a new species."

Not having 600,000 new species in the last 1 thousand years debunks the evolution misconception.

If we go back to the great extinction at the end of the Cretaceous, the number gets better, but not much. Assuming we start with 100 different animals and get 1 million in 120 thousand years we would double 14 times of double every 10 thousands years which also does not work.

Chromosome Dilemma Evidence

These chromosome things are neat, but they ALSO don't help us with evolution. As creatures evolve, increasing the genetic information contained in chromosomes enhances a species. This would be easily seen as an increase in chromosome packets; or so it would seem. It would be true if there were anything to evolution enhancement or uncontrolled evolution, but alas, no such thing has been found. Below is a short list of common animal types. Beside each animal type is the number of chromosomes used as the building instructions. Notice that "Man" is much more highly evolved than most of the animals as it has more instructions. Wow! The theory works. Man is better than other animals because it is more highly evolved.

Virus	1	Ant	2
Parasite worm	2	Indian deer	6
Fruit fly	8	Mustard	10
Micro-roundworm	12	Rye	14
Guinea Pig	16	Dove	16
Corn	20	Horsetail plant	21
Opossum	22	Kidney bean	22
Redwood tree	22	Chinese deer	23
Earthworm	32	Yeast	32
Frog	36	Pig	40
Mouse	40	Wheat	42
Bat	44	Man	46
Tobacco	48	Apes	48
Sheep	54	Domestic Horse	64
Wild Horse	66	Dog	78
Chicken	78	Carp	104
Crayfish	200	Fern	500
Butterfly	380		

Ape Chromosome Count

Oh! No! It seems that we have de-evolved from the Apes by this logic. Apes, however, have one more pair of chromosomes because two sets of pairs; those called 2p and 2q, are put together in the human set as one chromosome pair, so the theory still holds as our DNA information is actually more compact in humans than in apes.

Apes and the Backward DNA

It should be noted here that almost all of the chromosomes are identical when comparing human and ape sets, so we possibly evolved from them. The only chromosome packets that differ are the 4th and 17th set. These two also are almost identical, but appear to be inverted such that the sequencing is the same but split in the middle and recombined in reverse order. So an ape is simply an "accidentally backward human" or the reverse with a man being an **"accidentally backward Ape"**. While no one knows which came first; the man or the Ape, the evolutionist decided that humans can speak so they must have evolved from Apes rather than the other way around.

Horses

A little more DNA are noticed in a horse. That by itself seems strange, but even stranger is the fact that domestic horses have fewer Chromosomes than domesticated. It's like domesticating a horse not only changed its chromosome structure, but it made it less advanced as well. It probably would have been better for the horse if we never did such a dastardly deed.

Butterfly Masters

Now we continue down the list and find we will have to continue evolving for some time to get up to the complexity of a dog, carp, or butterfly. We should either respect our master butterflies or disregard this crazy notion of advancement by evolution. Those flying insects know that they are better than us; we don't have to acknowledge it to make it true. Chromosomes acknowledge it for us.

Archaeopteryx Dilemma

I know some of the "creationist theory advocates" are immediately thinking, "We have him now!" Creationist experts insist that the Archaeopteryx skeleton was forged to show a lizard with feathers. In fact, there are strong beliefs that many of the findings of ancient creatures are forged.

I'm telling you now that those assumptions are usually not true. There have been many mistakes concerning how skeletons are put together and an "occasional" forgery which tends to make some think that all findings showing genus crossover characteristics are forgeries that are manufactured as part of a sinister plot to turn evolution into a LAW. Guess what we found with the archaeopteryx? We found 7 almost identical species. Five of the fossil sets still had feather imprints intact. It may be remotely possible that some secret and powerful organization forged them all and placed them back into the ground for various researchers to find but that theory is really stretching credibility to the extreme limits. Following are an image of what this animal looked like and two different fossils from two different museums. These dinosaurs had wings; part lizard; part bird.

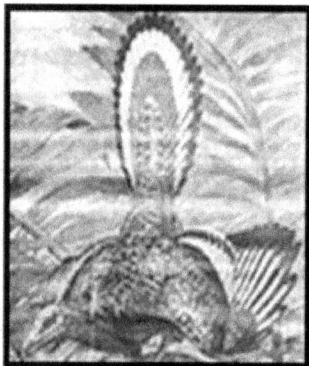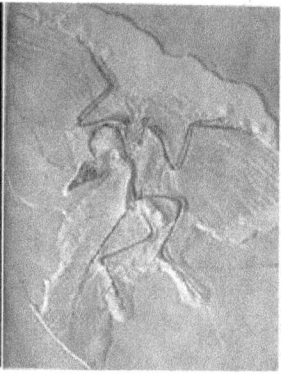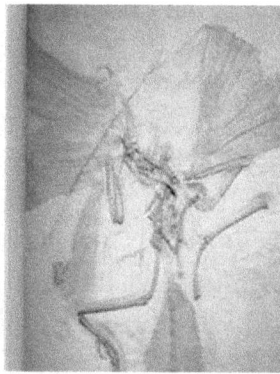

Even though this animal has both reptile and bird features, it does not prove evolution nor is it a survivable link species. This strange creature is a small part of the proof <u>against</u> evolution rather than for it. Where the evolutionist indicates that there would be survival of the fittest, this creature could not fly, but had these pesky feathers. They would not have increased their survivability with these feathers. Besides, this animal is only found in and around Germany which would limit any evolution opportunity. [Later we will find out that archaeopteryx could fly, because the earth was much different when he was flying around.] Trying to convince us that those wings were inappropriately designed doesn't hold water. Below is a blow up of still a third specimen of the Archaeopteryx from a third museum. The last one not only is for completeness, but also to show improbability of forgery.

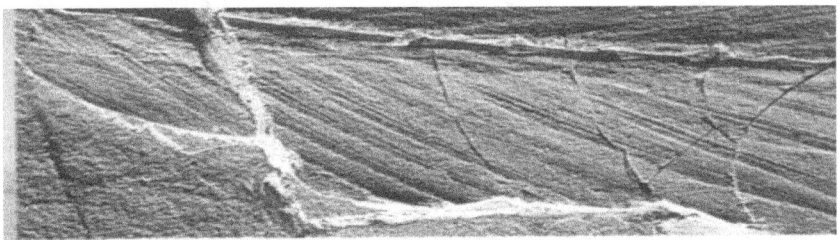

Note the fine detail of the feathers in the three archaeopteryx fossils shown. The first is from Berlin's Natural History Museum. The second was found in Bavaria, Germany and the third is a specimen from the Humboldt Museum.

Apes in the Wrong Place

While you're at it check out this set of prints for forgery. Why does this ape [next left] seem to have such a horrible look on its face? The reason is obvious to a casual observer, but hard to accept by the ardent evolutionist, because, the fear experienced by this mammal was because he was being eaten by an Allosaurus well before it would have become evolved to the high form of the ape. The find was in upper New Mexico, but typically you don't see these things in text books. One of the photographs of this impossible scene is shown next right and, no, it was not produced with trick photography. Since this find, more out of place dinosaurs like T-Rex, Hydrosaurus, and Stegosaurus have all been found with unfossilized bones meaning they were killed less than 50 thousand years ago, but that is another story. My little brain is thinking you can't fake soft tissue in a bone.

Mass Extinction Evidence

If we didn't have extinction periods, one might be able to establish some type of evolutionary timeline that made sense, but the earth DID go through some really awful times. Most people know about the 9 major mass extinctions. They were discussed in High School classrooms as the story of the Dinosaur demise was brought out. Almost every time, over 80 percent of **all** species on the Earth completely disappeared. We just accepted that this was possible in school because we were told that scientists had determined that this supposition was ok. Below is what we normally think of as the Geological timeline. It conforms to what we have typically been told and it is a useful beginning. What caused the major changes, what were the strange results that are identified by the chart, and how in the world could most of the animals disappear and miraculously reappear at least after each of the nine times we were told about? If you are comfortable just knowing that it happened, you don't need to read further, but to the right has a corrected time. Just close the book and be happy; but if you want to know a probable reason for the extinctions and replenishment of animals read on.

Standard Geological Timeline

Era/Period/Epoch	Time (M yrs. ago)	Time (T yrs. ago)
Archaeozoic Period	5000-1500	50,000-3000
Proterozoic Period	1500-545	3000-1000
Cambrian period	550-500	1000-900
Ordovician period	500-440	900-800
Silurian period	440-410	800-700
Devonian period	410-365	700-600
Carboniferous	365-300	600-500
Permian period	300-250	500-400
Triassic period	250-212	400-300
Jurassic period	212-145	300-200
Cretaceous period	145-65	200-100
Tertiary period	65-1.8	100-40
Pleistocene period	1.8-0.01	40-10
Holocene period	0.01-0	10-00

Less Know Extinction Evidence

We now know the timelines should be greatly compressed over the predecessor nuclear decay timing characteristics, but what I want to say even shows up with these massive timelines. What is not commonly discussed is that after each extinction boundary a large mass of animal types quickly reclaimed their former status and almost always exceeded the number of different animal types existing previously. In addition to that seemingly impossible fact is still another. Those 9 were not the only mass extinctions, which occurred. They were just the worst.

If natural genetic mutation and survival of the fittest wasn't the answer to this wide explosion of animal types after each extinction period, another theory must be introduced. If we accept, for a minute, the idea that human beings that "lived" for many thousands of years could have developed the capability to manipulate genetics and store "seed genetic material", it could be possible to quickly re-generate animals after each of the catastrophes. We will discuss this seemingly wild possibility along with others to test reason and consistency of the evidence rather than having "No" acceptable answer and ignoring possibilities, as some scientists seem to revert to when faced with this type of situation. Look at the following graph and note the extremely fast recovery periods that happened between each of the destruction events. The graph [top line] shows the pattern of extinctions that have occurred on the Earth over the past 600 million years [by nuclear decay that is now thought to have been less than 800 thousands years but Ice core, Hot-spot track, and Archeo-magnetic comparisons.] The bottom line shows what one might expect if animal differentiation and addition had occurred without "help".

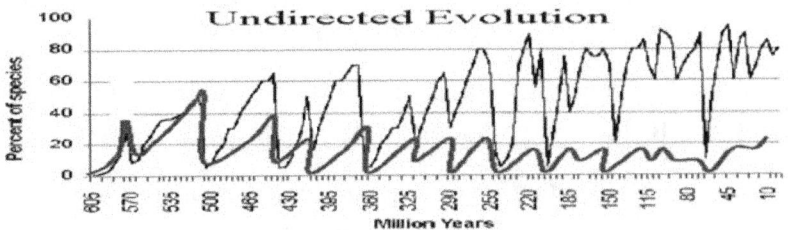

In order to emphasize the nature of the graph, the following chart was generated from basic extinction numbers determined at each of the major extinction layers. At each extinction period, I have provided the animal diversity just before the extinction as compared with the animal diversity of today. From it [Animal Diversity] we can determine that 500 million years ago [700 thousand with new timing] there were about 40 to 50 percent as many different species as we believe are in existence today and that diversity level generally increases between then and now along a slope that almost ignores the extinction periods. What should be noted is the extremely fast rise in animal life after each extinction period. Certainly, the fast rise was not possible if evolution caused the increase, because there just weren't enough animals that survived the extinctions to establish critical colonies.

Years ago	Boundary	Diversity	Animal Loss
1,000,000	Cambrian	35%	70%
900,000	Ordovician	50%	80%
800,000	Silurian	65%	90%
700,000	Devonian	50%	75%
600,000	Carboniferous	70%	80%
550,000	unnamed	50%	50%
500,000	Permian	65%	40%
430,000	Triassic	80%	95%
350,000	Jurassic	90%	90%
200,000	unnamed	75%	40%
250,000	Cretaceous	80%	70%
200,000	unnamed	85%	30%
180,000	unnamed	90%	40%
120,000	Paleocene	90%	85%
40,000	Pleistocene	100%	40%
11,000	Venusian *	110%	33%
10,000	Holocene	100%	80%

* [My Name]Biblical and other ancient Jewish works indicate that 1/3 of all life-forms died during this time.

The various Cambrian, Jurassic, Cretaceous and other eras essentially **exploded on the scene with animal diversity**.

A paleontological fact of life is that all known body plans (phyla) seem to have evolved suddenly.

Evolutionists are understandably uncomfortable with such a high rate of evolutionary innovation. Nothing like these diversity explosions has occurred in recent times and so rapid was speciation during the Cambrian Explosion that doubt is cast upon the accepted mechanisms of evolution which are slow, stepwise accumulation of mutations plus natural selection. Now multiply the speed by 800 times to correct the dating.

DNA Anomaly Evidence

To make the mystery of "**evolution**" even stranger, Dr. Wray and colleagues have analyzed the DNA sequences of seven genes found in living animals. Assuming that these genes mutate at constant rates and working backwards in time, they calculated that animal diversification would have had to begin about 1 billion years ago and this is assuming that no major extinctions occurred in between. I would believe that this same experiment could be reenacted over and over again with the same results. The only way for the diversification in a "natural selection world would be to allow for speciation over a billion years. The billion year expansion of the time frame gives accepted evolutionary processes the time it would need to innovate and create all those new body plans, but the initial explosion could not have occurred 1 billion years ago. To top it off, the earth had these "extinction periods" which would have greatly slowed down the diversification a lot. I know you weren't told about this anomaly in school, but the whole concept of how the animals came about and recovered is one topic that will be addressed in this book. Before we do that, let me make one thing perfectly clear. "God didn't make all the animals." After saying such a terrible thing let me say that the animals were made as a result of controlled evolution. Some call this intelligent design and immediately think it means God zapped the animals here.

"Controlled" Evolution

Let me start by saying God did zap some animals here. By almost all ancient records this is a fact and no science has uncovered a flaw in the statement. I'll explain a little more about this statement later, but note that I first said God did not create all the animals.

While the natural selection hypothesis, and the Creationist Hypothesis are both terribly flawed, this theory provides a better relationship between evolution model and *genetic mutation and manipulation*. It eliminates many anomalies IGNORED by many so-called scientists and religious leaders.

Implication of Genetic Breeding

Much of the evidence associated with the development of animals on this planet show something quite different than evolution. What the evidence shows is that some "group" was probably experimenting with genetics willy-nilly. They didn't seem to care what sort of oddball creature they made, provided that they could modify DNA and gain knowledge from the modification. The animals suffered and eventually the geneticists suffered. Some are thinking that this is absurd, but that is EXACTLY what we are doing today. We have "created" animals that eat oil, some that have wings coming out where eyes should have been, chickens with no feathers, and all types of "disease-animals".

In Italy, scientists have learned how to change the chromosome pattern in a pig so that a human heart will be generated instead of its normal heart.

In the UK they are genetically mutating the brain of a rat to make him smart.

According to MANY, MANY ancient books and stories, people did the same thing during the ancient days. The evidence suggests that genetic manipulation was going on but on a much larger scale than today.

God Didn't Make the Animals

I'm sure many people hate the statements I just made more than the "evolution" theory, but you have to look at all the mistakes.

- *The useless, long, wrapped around nose, of the parasaurolophus mentioned previously*

- *The arms that are too short for the Tyrannosaurus,*

- *The requirement for giant squid to lose their penis when trying to procreate, [Giant Squid penises are continuously being found on the ocean floor as they, many times, missed their targets.]*

- *The exploding butt of the bombardier beetle.[Flames leap out its butt for protection and people think it is so very cute.]*

- *The blood spitting eyes of the Texas Horned Lizard [What a neat way to use eyeballs.]*

- *The loss of the penis bone in humans, while almost every other mammal has this strange characteristic. [Monkeys have them and most other mammals but the hyena and man are "boneless"—Go figure? [What does go figure mean?]*

- *The taste buds on a butterfly's legs while most animals taste with their mouths.*

- *The hearing nodules on a grasshopper's leg while most animals use ears. [**This reminds me of an experiment done by geneticists recently where they produced a fruit fly which had legs develop where its eyes should have developed. It seems irresponsible, but geneticists were proud of themselves.**]*

- *The requirement for a frog to push his eye inward to help him swallow a meal. [**It may be where the term "his eyes were bigger than his stomach" came from, but totally absurd.**]*

I could go on and on with ABSURD, inappropriate "Experiments" done on animals in an attempt to bring out a specific characteristic or another. That brings us to the most obvious statements about the vast number of inappropriate animal changes.

GOD would not have "experimented with size, characteristic, speed, eating methods and other items that did not foster an increase in civilization or survival. Another must have "helped".

Some say that this obvious "experimenting" with animals proves that there is NO GOD. **Well, they are probably are right in the first part of their statement and totally wrong in their determination.** God would not have done the experimenting, but the evidence suggests that the only reasonable conclusion for creation is that a single creator [God] must have initially designed the animals and all other life-forms.

If God wouldn't do all this experimenting and Evolution would not allow it, something else is most likely the answer.

Clean Evidence

Clean may be a part of the answer. Have you ever wondered why the Bible talks about "clean" animals and unclean/abominable animals? Almost ALL animals recognized in the Bible are in the abomination side. Noah was ordered to take many of the abominations with him during the worldwide flood thing, but the Jewish people weren't even allowed to eat the nasty things. The list below is not a complete list of the separate types, but has been presented to show a point.

Type	Clean or "good" animals	Abominable animals
Bird	Pigeon, Robin, Duck, Dove	Pelican, Ostrich, Eagle, Swan, Owl
Mammal	Horse, Cow, Goat, Sheep, Donkey	Camel, Pig, Monkey, Ape, Porpoise, Whale
Reptile	No clean reptiles	All
Amphibian	No clean amphibians	All
Fish	Only those with Fins and Scales	Squid, Eel, Catfish
Insect	Locust, beetle	All Others That Fly and all that don't

This is not a list of things good or bad to eat, nor was it a list of what would be safe for Adamics to eat with respect to disease. Certainly an eagle and a swan are more majestic than a pigeon and a duck and to think of a porpoise, whale, and monkey as abominations seems strange at first.

What Makes and Animal and Abomination?

What this list seems to be and evidence supports, is a list of animals that had been mutated by the ancient geneticists that EXPERIMENTED before the flood. Before you say I'm crazy again, come up with a more plausible answer why monkeys and dolphins were abominations. Apparently, the genetic codes of these animals were so messed up that they were abominable to God according to the Bible. Other explanations don't seem to fit nearly as well, so I'm sticking with my hypothesis.

"Unclean" meant "mutated" animals. The animals were mutated by humans not by GOD.

Noah brought "God made" animals and "mutated" animals on his Arc. But he did NOT bring 3 million types of animals on his small boat. Later, the surviving animals were "mutated" some more by the post flood geneticists. Don't slam the book shut, relax and read the evidence.

God Warning Evidence

God abandoned the "unclean" animals and warned against making more. All through the Bible there is a major separation between the two types of animals and other works give us more specific reasons for the separation.

Excerpts from the "Book of Giants"

This is an Essene text which was transcribed many times and well recognized as a book of authority during ancient times. It not only tells about giant humans in the past, but also describes problems associated with genetic manipulation.

[The rulers selected two hundred] donkeys, two hundred asses, two hundred [rams of the] flock, two hundred goats, two hundred [other beasts of the] field. From every animal, and from every [type of human was taken its seed] for mixed sex. [After a time] they defiled [the animals and people and begot] giants, monsters, and dragons. [It strongly suggests that human and animal genes were mixed together to produce new animal types.]

Excerpts from the Book of "Jasher"

This book is considered canon by some Christians and is referenced in several books of the Bible as a great teaching tool. It tells us the same thing.

*Jasher 4:16-18- And all the sons of men taught one another their evil practices and they corrupted the Earth, and the Earth was filled with violence. And their rulers went to the daughters of **men*** and took [*this taking is sexual*] *their wives by force from their husbands and the sons of men in those days took* [*not sexual usually*] *from the cattle of the Earth, the beasts of the field and the fowls of the air, and* **taught the mixture of animals of one species with the other**, *in order therewith to provoke the Lord; and God saw the whole Earth and it was corrupt, for all flesh had corrupted its ways upon Earth, all men and all animals.* **[Corrupted animals did not mean the animals were evil, it meant that the species were changed inappropriately.]**

Excerpts from the Book "Enoch"

This book is also considered canon by some Christians and is also referenced in several books of the Bible as a great teaching tool. Like the book of Jasher, it tells us the same thing.

Enoch 7:5- *And they began to sin with birds and with animals and with reptiles, and with fish.* **[This did not mean that the rulers had sex with fish. This is talking about manipulation of species]**

Excerpt from the Book of "Jubilees"

This book is also considered canon in some religions and was a respected ancient book during the time that the Bible was being written.

Jubilees 5:2- *And lawlessness increased on the Earth and all flesh corrupted its way, alike men and cattle and beasts and birds and everything that walks the Earth all* <u>corrupted their ways and their orders.</u> **[The only way that animals corrupted their way was that they were genetically manipulated and just weren't the same animals.]**

Jubilees 7:24- *Afterwards* **they sinned against beasts** *and birds and everything that moves or walks upon the Earth. [***There are two ways to sin against beasts- sex and genetic manipulation. God didn't like either.]**

Experts from the Book of "Enoch II"

This book is canon in some Bibles and it also tells us the same thing.

Enoch II 59:5-6- But whosoever kills a beast without wounds, kills his own soul and defiles his flesh. And he who does any beast any injury whatsoever, in secret, it is evil practice, and he defiles his own soul. **[The killing and injury done in secret was not killing animals for food, it was the genetic manipulation and corruption by integrating man's; genetic material.]**

Excerpt from the Book "Generation of Adam"

While this is an ancient Gnostic book, and not a generally recognized ancient Biblical text it tells us the same thing.

Generation of Adam 6:1-5-Among our little ones was Timnor and Ammah. Timnor understood physical law and created mighty machines. Ammah understood the secrets of creation. She manipulated the very fountain of life until she had created new forms of beings dedicated to the destruction of mankind **[As in other texts, manipulation of species was common practice and the results were not always helpful to man and were always against God. In this case, the verse is talking about the direct descendants of Adam manipulating animals.]**

Excerpt from the book of "Naphtali"

While this is also an ancient Gnostic book and not a generally recognized ancient Biblical text, it tells us the same thing.

Naphtali 1:25-27-   The Gentiles went astray, and forsook the Lord and changed their order, and obeyed stocks and stones, spirits of deceit—become not as Sodom, which changed the order of nature. In like manner the watchers also changed the order of nature, whom the Lord cursed at the flood, on account he made the Earth without inhabitants and fruitless. **[Not only does it indicate that the watchers and gentiles [not pure Adamics] practiced genetic manipulation, but also that the practice was the major cause for the flood.]**

Excerpt for the Book of Zadspram

While this is an ancient Zoroastrian book, it also tells us the same thing.

Zadspram [Iran]- *From the seed which was the ox's, they would carry off from it and the brilliance was entrusted to the angel of the moon in a place that seed was thoroughly purified by the light and was restored in its many qualities.* **[After the rulers corrupted the animal genetic code, God had to recreate them.]**

The Zoroastrian "ZAND-AKASIH" *- Satan miscreated creatures and they became useless. God saw the defiled and bad creatures, they did not delight Him* **[They became abominable]**. *Satan's downfall was the unrighteous creation of the creatures and ignorance.*

I think you can get the picture. According to all ancient Middle Eastern history that describes the problem, most animals were DESGINED by people. You can believe it or not. For you to believe it, you must first believe that people had to be here to design the animals millions of years ago.

The idea that advanced humans were here before the "lower" animals sort of messes up "survival- of- the- fittest evolution" but that doesn't mean people weren't here. It simply means the evolution idea must change. Advanced humans were "created" out of sequence by an out of sequence generating God.

Don't whine about the information and keep it away from our children accept it and accept "Intelligent Design Evolution" as a much more LOGICAL and much more FITTING theory for interpretation of the evidence. The evidence tells us about ancient humans.

Ancient Humans

There is an enormous amount of information and evidence concerning an ancient civilized race of humans. While I'm not going to get into the large amount of written texts which confirm very ancient humans, I think it is important to understand that they were here to help mold our ancient planet. This seemingly unbelievable element in the development of the earth is not a fairy tale nor is it a lie. Evidence shows that modern humans were here during the Triassic and Jurassic periods. The evidence of the societies, the science, the capabilities, the wars, and the disappearance of this first race of people is extremely interesting, almost unbelievable, and proven in the stone.

The Greeks and other groups around the world termed this ancient time as "the Golden Age and 1st of 3 creations of humans". Ancient stories about humans are almost identical. Not only did this "Golden Age" include civilized humans living on earth, but also there is strong evidence to suggest that they were on other planets, as well. I touched on some of that evidence earlier. The near planets were inhabitable and evidence suggests that people lived there. The people on earth knew about flying, the planets, genetics and many other things and "stone evidence" is about the only thing that can survive such a long history.

The Sumerians told us that the first inhabitants changed an animal into a human and his hair fell off his body. Others tell an almost identical story. The Bible may even tell us details about the ancient humans and their genetic science. We know that the Earth was populated before the Biblical "6th day or age" development of human beings. We know that the "6th day or age" was the re-creation of "humans" because we have found many ancient writings, physical evidence, and the Bible confirms it in the 1st chapter of Genesis.

Genesis 1:24-30-And **they** *said, Let* **us** *make man in our image [on the 6th day]. So* **the Elohiym** *created man in their own image. The Elohiym said unto them.* **Re-plenish** *the Earth,* *and subdue it:* [Repopulation would only be necessary if a first civilized population had generally disappeared during an ancient time.]

An ancient human lived during much of the "evolution of new creatures" and was very civilized. We will look at a small amount of the enormous quantity of evidence that has been obtained over the years. Not only have we found many drawings of people and dinosaurs together, but also we found that humans stepped on small dinosaurs and walked in the footprints and possibly even became part of the footprints of much larger ones. After a long time, the dinosaurs became the mightiest animals on the Earth. The original humans had almost completely disappeared, over time, but it took hundreds of thousands of years.

Elohiym

Some probably noticed I didn't say "our creator God" made man on the sixth day because the Bible doesn't say that. The writer of Genesis made a point of NOT saying that. He even used words like "us" and "they" to make sure there was no confusion. No! The verse was absolutely not talking about "Us the trinity of God", as he is not referred that way anywhere else in Genesis that way. There are two specifically identified "gods or masters" identified in the first 2 chapters of Genesis. One is simply called "Elohiym" and the other is called the "Lord over the Elohiym". There is a reason for the two names and it has nothing to do with some exotic "Interpretation". If there are 2 called out, it is talking about 2 different entities. One should consider that the "Lord over the Elohiym" is the true creator; not only Lord over the Elohiym, but Lord over everything. The word "Elohiym"[which is the plural form of Ela or god] does not mean that there were many "Creator Gods" as some have suggested. To the writer of Genesis it seems that they were the "masters that knew about genetics". If we assume people had been on the earth for a very long time, it is not

105

unreasonable to believe that they were "helping" in the manipulation of genetic code structure and made a version of human being. Elohiym seems to be a reference to the remaining humans that used to inhabit the earth before the REPOPULATION was required.

By the way, this 6th day human was not the Adam that was created by the "Lord of the Elohiym" in chapter 2 of Genesis.

Just like the Greeks had indicated and just like the Maya had indicated there were three distinct creations of humans and Genesis "generally" confirms all three. The Greeks indicated that the 1st "creation" was the Golden Age. It would have lasted from the Jurassic until less than a million years ago by many accounts. Not only does the evidence suggest that these ancient humans lived over that time period, but they also got pretty smart over that time period and they worshiped God as their creator. If you wonder why we have such a large brain, one answer might be that the ancient humans used to use that big old thing and were able to do things we still can't do today. One of the things they could do involved animals.

Genetic Manipulation

Well before the Neanderthal, evidence suggests that "Ancient Humans" were very similar to the humans of today. They still got old and died like we do today and they probably thought that it would be a good thing to manufacture spare organs just like we are trying to do today. By making new parts and other changes, their life-spans began to extend well beyond that of modern humans.

Finally the evidence tells us that they got caught up in the designing of strange animals and they got very good at the whole genetic manipulation thing. They started small but soon they were making all types of animals. Every once in a while, a meteor or similar object would hit the earth and destroy their work, but they would just start over and make some more animals. I know some will scoff at the concept of ancient humans being real, but there is evidence. *A Lot of Evidence!*

Evidence of Ancient Humans

Very ancient humans left their mark on the world over well over a million years ago. The creatures in the ancient communities looked like us, walked like us, made things like us, but were here before the time that God created the "sixth day man" described in our Bible. Don't take my word for it, look at and read about some of the collected evidence. The time that the ancient humans were the main physical creature started well over a million years ago and lasted up until about ½ million years ago. This was during the entire known "evolutionary period" of the planet so it is pretty important as we try to piece together the earth's development. What we know about this human is mostly in the form of physical evidence. From this evidence, we know that they were here, they were civilized, and they were abundant. Some of the ancient texts referenced and confirmed their existence, as well.

Even though we have a large amount of physical evidence that this human left behind, scientists are uncomfortable with these artifacts because they generally disprove their other theories. Those holding true to a 6 thousand year old earth are uncomfortable as well, because the ancient humans disprove many of their statements. Don't let people just tell you that the dates are all wrong. If you find objects buried beneath ½ mile of coal; that coal did not pile on top of itself over and over again within a couple of thousand years and crystals did not miraculously appear in a matter of days. Most of these items and the general dating of the artifacts that make up much of the ancient human evidence is real and the articles found are extremely old.

Physical Evidence

Archeological evidence on Earth shows some interesting clues concerning life during a very, very, ancient time. We can get a very reasonable idea about the age of things found on the Earth, because of the long list of dating methods at our disposal as discussed in a previous chapter. The time began before the age of the dinosaurs and during the time when the huge, coal producing, forests were still here. At that time the forests had not been compressed into coal. After many years that was changing. Along with the remains of the trees were caught remains of a civilization that began when the coal forests were still trees. Bits and pieces of materials were caught inside the forming coal and shale beds that are now being harvested. Continuously, the signatures of the very ancient dwellers have been found. The evidence strongly suggests that a large number of humans lived on the Earth during this time and they stayed for a very long time.

This evidence wasn't written down so "opinion" has not tainted its true meaning. This evidence is in the rocks themselves and, I believe you could say that "Rocks Don't Typically Lie".

Most of the descriptions below come from multiple sources; many of which are provided in the bibliography. Please investigate these things further by reading additional accounts. I have purposefully made the descriptions short to introduce them to as many people as will care to look.

Shoe Evidence

Many footprints have been found along with bones and other remains of what could not have been here unless humans inhabited the world. Here are a few of the many examples found so far. If there were feet, then humans were not far behind. One of the questions will be, "Why are they barefooted if they were civilized?" The answer, I believe, is that impressions of footprints can only be made around areas like the seashore where the impressions could be quickly filled with other deposits to preserve the image. Why would they use shoes at the beach? The first image shows where a person was walking in between dinosaur footprints [barefooted]. Even so, sometimes shoe prints have been found.

Nevada 1927-A sandal sole print, shown middle below, was found in a coal seam. Even the impressions of the threading holding the shoe together could be discerned. Estimated age was Jurassic.

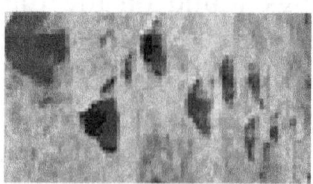 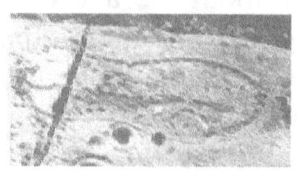 

Washington State- The photo on the right was taken in northern Washington State and was reportedly found with another partial imprint. It is the 16-inch long shoe print of a large individual. The rock itself was determined to be Tertiary.

Utah-1968-The picture following is of a fossilized, 10-½ inch long human sandal print found next to a smaller human footprint. Live trilobites were crushed by the sandal in the same stone before fossilization. Columbia Union College made studies on the fossil an attested to its authenticity. The age is during the Triassic or Jurassic period. The bumps near the heel and another near the toe are the locations of trilobite carcasses.

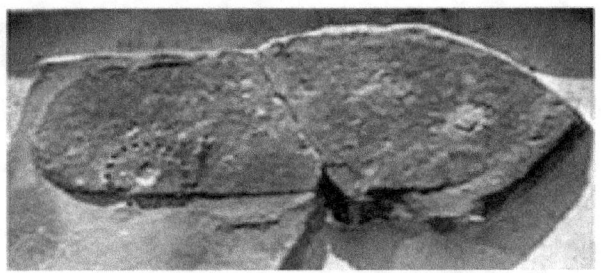

Utah 1969-More footprints found near Salt Lake City- 2 more sandal prints were found with trilobite fossils nearby. The estimated age is Jurassic. [There must have been a large settlement in the area for so many footprints to be found.]

Dinosaurs and Humans Together

Much evidence has been found showing even the early Trilobites were coexisting with the ancient humans. These trilobites probably died during Jurassic times and men stepped on them before they became extinct. Just think about it!

The strange part of this information is not that humans were here so long ago, but that they survived so very long through all types of disasters on the earth.

Huge Men and huge dinosaurs also walked together. Some scientists use that finding to "prove" that dinosaurs were here only 6 thousand years ago. I believe I went over many of the studies which continuously show that coexistent footprints don't mean that the dinosaurs were here a couple of thousand years ago, they mean that man was here long ago. Man didn't step on the huge dinosaurs as he had done to the trilobites, in fact, the humans might have been some ground cushion for the dinosaurs from time to time. Generally they coexisted.

Kentucky-1938-Three pairs of tracks [human] were sunk in gray sandstone [once a sandy beach during the Cretaceous period]. Photomicrograph studies showed that the tracks were not manufactured artificially or recently so don't start thinking fake feet.

Texas-Texas is full of sites- The one on the left has 14 human and 134 dinosaur tracks together and estimated to be Cretaceous.

 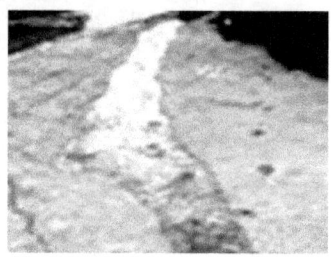

More Texas-Another Texas site to the right. This one had 15 human prints near dinosaur print finds and was estimated to be Cretaceous.

Texas 1976-Human and dinosaur prints were found together at Glen Rose, Texas. The dinosaur prints actually went over the top of one of the footprint impressions in a series. 203 dinosaur prints and 57 human prints have been found in the same area. The largest human print was over 16 inches long. Estimated age is Cretaceous. The one [next left] was found near the main group.

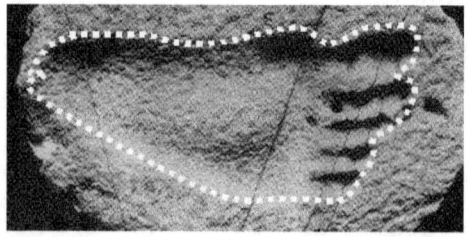 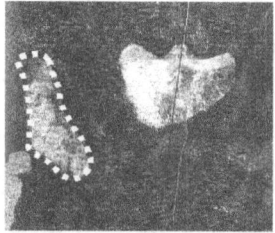

Texas 1971-Another example from Glen Rose above right shows a man going one way and a dinosaur going the other. Lucky for the man and the footprint is huge [18 inches long]. The man might have been as tall as the baby duckbill dinosaur going the other way. It is estimated age is Cretaceous.

Utah Example

Here are ones from Utah and Arizona. Can you make out the drawings? I drew it in about the actual drawing on the one from Utah to help. [See below left]

 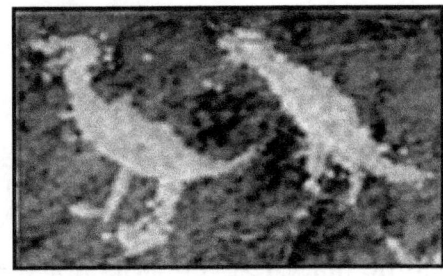

Arizona Example

In 1924 the Dohenny Scientific Expedition found Petroglyphs in the Havai Supai Canyon. Some of the pictures were of prehistoric beasts. One was of the Tyrannosaurus Rex originally believed to have been extinct since the Cretaceous ended. The pictures are fully patinated [a process of growing rust which typically takes many thousands of years], so the probability of recent artistry is extremely low. [See above right] From these and many more of the drawings found there can be little doubt that ancient inhabitants of North America lived with the dinosaurs. That doesn't mean that dinosaurs lived here a thousand years ago, it means that people could have lived here hundreds of thousands of years ago.

I'm not going to tell you my thought on this one, because we have to get back to a very important time in earth's development that was witnessed and possibly even helped by these civilized inhabitants. That event was the complete destruction of Venus as a livable planet. The reason Venus stuff is important to us is that earth was greatly affected. Not as affected as it was from close encounters with Mars, but it was not a pretty time in earth's history. Some believe that the worldwide flood that followed a few thousand years later was the most traumatic, but both events were doosys. We need to spend time on both and look at the evidence.

Venusian Catastrophe

Let me start off this section by giving a brief overview of some of the "passed down" information from eye witnesses. I'm talking about what the people living 10 thousand years ago saw. We might assume that the event was not well known during ancient times. We can only find written information about this event from Europe, South America, North America, Central America, Egypt, Sumeria and the rest of the Middle East, India, China, and the Pacific Islands so we shouldn't believe it happened. Some might tell you it was worldwide mass hysteria, but I have a hard time ignoring just about everyone writing about this time period. On the off chance that you might believe, here are a few of the written testimonies. What you will find is that there is a common theme.

The planet we call Venus used to look a lot larger in the earth's sky.

The planet looked comet-like during this time period. It had a tail that could be seen easily by all observers.

The planet had some major catastrophe on it which made it appear to explode

Pieces of debris from the explosion pelted much of the world.

Many places caught on fire and many people died as a result of this assault on earth.

Venus appeared to catch on fire as well and changed its position in the sky.

Written Evidence

The fact that there was a terrible Meteor storm witnessed on earth was captured in a few ancient histories. Here are some from

Europe, South America, and North America, all over the Middle East, the Far East and the Pacific Islands.

European Venusian Meteors

The European countries remembered the devastating event and wrote about it. Greek stories are filled with details. Here is one.

According to Greek legend, "A blazing star almost destroyed the world with fire before it became **Venus***."* [Although it is difficult to interpret, here is what I believe they are trying to say. I believe they are saying Venus used to look like it had a blazing tail, something happened to it which almost destroyed the earth and finally, it settled down to become the Venus we now see.]

South American Venusian Meteors

The People of South America remembered and wrote about it. The Inca legends tell the story.

The Inca called **Venus** *the "Wavy haired planet";* [This also seems difficult to interpret. Could wavy hair be flames shooting from its surface during a time when the Inca were around?]

Central American Venusian Meteors

The People of Central America remembered and wrote about it. This is from one of the Aztec legends.

The Aztecs called **Venus** *"the Star that smoked" and said that it once passed by the world blazing and killing many people. The Aztec god, Quetzalcoatl, associated with Venus, is typically pictured with a wavy headdress.* [I'm going to get into this whole wavy haired tail thing in a minute, but the blazing and killing sounds like meteors hitting and setting fires. The meteors came from the smoking planet.]

In the Mayan Dresden Codex*, the god of Venus is depicted with shooting darts. It seems to me that if something is shot away from a planet, it would have been meteor-like. The picture below is from one of the Dresden pages. You guessed it Venus is in the middle.*

Black foot Indian Venusian Meteors

The People of North America remembered and wrote about it. Let's see what the Blackfoot had to say.

*According to their traditions, "The morning star [Venus] put on a scarlet cloak [**sounds like it was on fire**.] And appeared before a woman on Earth that he loved. She went into the sky with him, but was warned never to look back. She did, of course, and was ordered to return to Earth."* [The return was a mess if we believe the other histories.]

Ute Indian Venusian Meteors

The Ute Indians tell us the same thing in their verbal history.

The sun was slivered into a thousand fragments, which fell to Earth causing a general fire. Then Ta-wats fled before the destruction he had wrought. All were consumed; until at last, swollen with heat, the eyes of the god burst and tears gushed forth in a flood which spread over the Earth and extinguished the fire." [This flood is not the worldwide flood we have all heard about, but it was significant, just the same. As far as the sun bursting, I personally believe it was Venus and not the sun.]

Egyptian Venusian Meteors

In Egypt, the event was known and written about.

Sonchie, the high priest, told Solon, a Greek historian, about events before the flood. He wrote, "Many are the destructions of mankind that have been and shall be. The greatest are by fire and water. During long intervals there are deviations of the bodies that move around the Earth in the heavens and the consequence is widespread destruction by fire of things on the Earth." *[The fires must have been everywhere when the Venusian moon split apart. The comment that it was one of the "Normal bodies that moved around in the Earth sky" limits the body to one of the close planets. Of course, the closest is Venus.]*

Jewish Venusian Meteors

The Jews wrote about the event in the book of Enoch chapter 85 verses 1-4 we read:

"A single star fell from heaven- raised up and fed among the cows--I saw many stars which descended and projected themselves from heaven to where the first star was." [Some claim this and similar verses are figurative and depict Satan being thrown from heaven, but sometimes people simply write what they want people to read.]

Sumerian Venusian Meteors

The Sumerians made record of the blazing tail of Venus. Their goddess named Inanna was associated with Venus and the information is the same as recorded by all the rest.

"To the queen of the heavens Inanna [Venus], to her who filled the sky with her pure blaze. The luminations are as bright as the sun. Who initiated the flood-storm? You roared in the heavens and Earth. You smote the flesh of the people." [The blaze of Venus filled the sky, roared across Earth and smote the people. I think the only way Venus could smote the people is if its moon exploded and pieces fell to earth as a huge meteorite storm.]

"She [Inanna/Venus] who causes the heavens to rumble. She who shakes the Earthquake. She cried toward heaven and Earth, "My hair will whirl in heaven for you." You flash like lightning over the highlands. You throw firebrands across the Earth. You split apart the mountains. [The hair extending sounds like a reference to a comet-like tail or a blasted away section of Venus that hit the Earth. Firebrands hitting the earth sounds like meteors to me. Your firebrands might be different.]

Assyrian Venusian Meteors

115

Assyrian literature tells the same story. This time the Venus-goddess is named Ishtar, but it is the same.

"To the pure flame that fills the heaven, who shines like the sun 'Ishtar" [Venus]—"I ran battle down like flames in the fighting. I make heaven and Earth shake. I trample the Earth. I destroy what remains of the inhabited world". **[To destroy the remains of the inhabited world, there must have been something substantial that happened with Venus.]**

Arabian Venusian Meteors

Coptic texts date the event for us in the Age of Leo. The ancient Arabic text called "Bundahishn" tells us the following:

The Ancient Coptic text tells about a great fire and flood coming out of the constellation of Leo. [This not only describes the event but places it in the "Age of Leo", 11 to 13 thousand years ago.] *It goes farther indicating that the beginning of world history was around 10 thousand years ago and some of the major deities were born during this event.* [The beginning of history must have meant that there was a destruction period just before that time.]

Phoenician Venusian Meteors

Phoenician texts describe the event, but this time the Venus-goddess is Astarte, the Phoenician version of Ishtar.

"See, Astarte" [Venus], she descends into a pool as a fiery falling star". ***[A beautiful description for a meteoric terrible disaster.]***

Persian Venusian Meteors

Mandean Texts from Persia give us the same information.

"150 thousand years after man was created, the whole Earth broke out into flames and only 2 escaped." It continues by saying that they had children and, of those ancestors, Noh [almost like Noah] was the one that survived the Flood that followed. [The Earth being filled with flames could have been from the huge quantity of meteors from the explosion, but clearly this event occurred well before Noh survived a worldwide flood.]

Indian Venusian Meteors

The Indian writers also informed us of this terrible calamity. The people remembered Venus sweeping away the stars.

*Indian literature states the following, "Her [Venus's] anger grew so terrible that she transformed herself, grew smaller and black. On a blind rampage she was killing everything and everyone in sight. **Her hair is wild**, her eyes red. The world trembles and cracks under her tread. Her dark hair flies in the sky sweeping away the sun and stars."* [Again we read about the comet-like tail and so many meteors that the sky is darkened.]

Chinese Venusian Meteors

The Far East writers also informed us of this terrible calamity. The people remembered Venus sending down a huge meteor shower.

The Chinese writers said the same thing, "There was a time when a planet [Venus] approached close to the Earth, causing great showers of stones." [Not too many of the planets could have come close to earth. The moon of Venus is my guess.]

Venus was depicted as a dangerous "fire spitting planet" according to Chinese legend.

Pacific Island Venusian Meteors

Even the people of the Pacific remembered Venus sending down a huge meteor shower.

Venus was depicted as a dangerous, "fire spitting, planet" by the Samoans. [It is like reading the Chinese version. What would have given them that idea?]

Physical Meteoric Evidence

Not only did people write about the event, but also, there is a huge amount of physical evidence in the form of thousands of craters left when the flaming masses hit the Earth. Pieces of material from the explosion hit places around the world. We know when they hit, we know that the explosion that caused them was reasonably close, and we know that there were many thousands of meteoric chunks that hit the earth at the same time. Here are some of the things we know

- *We know earth was peppered with thousands of meteors all at one time*

- *We know that the meteors struck 12 thousand years ago*

- *We know that the water temperature in the Atlantic and Antarctica both rose at that same time period*

- *We know that the Water level of the Atlantic ocean rose by a huge amount at this time*

- *We know that an Ice Age ended at this same time.*

- *We know that the earth's axis shifted at this time.*

- *We know that Venus's surface caught on fire at this time*

- *We know that a large quantity of huge craters are located along Venus's equator and nowhere else as is the meteor that struck was rotating around the planet.*

- *We know that hundreds of written texts indicate that Venus changed significantly in recent past.*

- *We know that there is a plasma string between Venus and Earth indicating that some electrical connection was strong in recent past.*

- *We have heard that great commerce center Island nations sank in the oceans about this same time.*

- *We know that this was not the worldwide flood addressed in the Bible.*

I know you haven't been told about this previously, but it happened just the same. Here are a few of the many pieces of unimpeachable evidence.

Worldwide Meteorite Evidence

Large amounts of "meteoritic mass" and an estimated 500 thousand strange indentions, strongly believed to be from massive meteorite showers have been found around the world that date to the end of the Pleistocene era, about **10 thousand years ago**. Large quantities have been found in United States East Coast, Alaska, Siberia,

118

Bolivia, and Netherlands. Guess what! The time period for the destruction of the Venusian moon is about 10 thousand years ago. If they both happened about the same time, there is a good possibility that they were the same event.

Glass & Stone Evidence

Tektites are small pieces of glass formed as a meteor strikes the ground and melts the surrounding area. Many have been found in sort of an "S" shape and distributed over large portions of the Earth.

Some were found embedded in fossilized wood, in Australia,

Others were found in Vietnam

Others were found in the Indian Ocean.

Several dating methods were used including Stratographic, Carbon 14 and others. They showed that most of these pieces were deposited around 10 thousand years ago. Ok! Maybe the ones inside the fossilized wood came from an earlier strike, but most were Pleistocene Era events just like the Venus moon blast.

New Zealand Evidence

Today, huge quantities of metallic meteorites as well as objects called "china stones" can be found everywhere on the island nation of New Zealand. Inside the stones are the remains of burned up **Pleistocene** type material, which dates the event to between **10 and 20 thousand years ago.** [I suppose you think these came from the Venus moon strike just like me.]

Carolina Bays Evidence

While the above examples show an unbelievable stress on the earth, the best examples of meteoric evidence can be found in the United States. The east coast of the United States was pelted with huge quantities of objects 10 thousand years ago. There are still an estimated 500 thousand meteorite indentions called "Carolina Bays", which mark this incredible event in history.

Let me wonder for a minute at an absurdity. One hundred and forty thousand of these 500,000 meteoric blast holes have

I'm not sure what is scarier, this terrible event or the fact that people EASILY ignore it simply because it makes them uncomfortable. Just think about how uncomfortable the people of that time were as they essentially saw the sky fall all around them. The picture below shows the major areas where these objects have been found in the United States. **These generally date around the same time. The evidence shows that the Venusian moon most likely met its end at the same time that these 500 thousand holes appeared and the other holes around the world described above.** Some of these indentions are very large and have diameters that are thousands of feet across. So it wasn't just a little meteorite storm.

Carolina Bay Craters [greater than 1000 feet across]	
State	Number of major Bays
Georgia	27
South Carolina	102
North Carolina	202
Virginia	17

The direction of the blast also gives us useful information about our developing planet. The clear indication tells us that the "10 thousand year ago equator" was in the direction of the meteorite path. The Carolina bay incident was a huge onslaught of meteors striking the Earth, which caused holes everywhere. Don't just take my word on this. Below is a picture showing the quantity of these things in a small area. To make them easier to see, I put rings around the larger ones. There are 25 or more in this area alone and they are found along almost the entire Eastern coast of the United States.

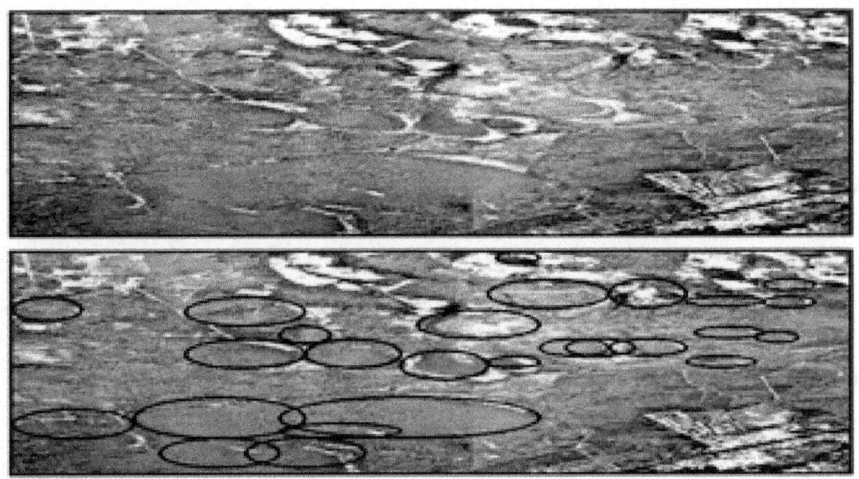

Just think about how it must have been that long time ago if you happened to live near South Carolina and literally hundreds of thousands of meteorites blasted the entire area over a period of perhaps 6 months. Most of your friends would have been killed and any civilization that was built up would have been in shambles. That is what the world would have been like, but the worst wasn't over—and for those who think that the Carolina Bays look like a multitude of sink holes caused by eroding caverns in an underground aquifer, think again. The underground area below these "Indentions" is not a limestone honeycomb and sink holes could not be the cause of the thousands and thousands of craters. These came from meteors and the meteors all came about the same time [10 thousand years ago]. The explosion that caused the meteors had to be fairly close astronomically speaking because the pattern is fairly confined along the path shown. This whole Carolina Bay thing is so revealing that it should be studied in our schools, because it shows us evidence of the earth shifting on its axis sometime after the bays were formed. It also establishes the cause for extinction at this time. It makes us look a Venus more closely.

Physical Evidence on Venus

Finding hundreds of craters around the world is one thing, but if we are to believe that Venus is the actual planet that caused wide destruction of the Carolina Bays, Venus would have had to have been closer to the earth. After the incident, Venus would have moved away from the Earth and have taken its new position in the solar system. It would have been similar to when Earth moved away from Mars, as discussed previously, only this time the movements would have occurred only 10 thousand years ago. With the Earth and Venus being close together, strong interactions could have come into play that eventually disrupted the "weaker" planet. Remember we already have two major elements of evidence. One is a large set of accounts from all over the world that directly indicate that huge meteors fell from Venus. The second is the fact that the surface of Venus erupted into an inferno very recently. Now scientists have obtained a third piece of evidence, which shows that Earth was involved in the destruction. The evidence is called "plasma trails".

Venusian Plasma Evidence

Outer space and our planet are both filled with something called plasma. Plasma is just a name given to a connected group of ionized atoms that can be influenced by magnetic fields. This plasma is strange stuff. It can even conduct electrical current. This current flow creates a magnetic field, which also affects the structure of the plasma. Scientists have not been able to adequately model this plasma phenomenon mathematically, but we have a great deal of empirical knowledge of plasmas, because we use plasmas every day. Arc lamps, arc welding, and even a neon light are all basically plasma generators.

Plasma trails can be and are produced from violent electrical disturbances occurring on planets, if they are significant enough to cause **quick atomic ionization**. This atomic ionization would be expected if parts of the planet were yanked away or hit with a huge electric field or hit by a huge lightning bolt. Additionally, we must understand that plasmas, although they are basically lumps of gas, do not behave like gases. They develop structure. When a huge variance in electric potential gets too high it produces a huge electric current that, in turn, causes the ionization. The current flow also causes these huge magnetic fields that, finally makes the something we call plasma filaments [**tails**] that twist together into things that look like "gas ropes". As long as the current continues, the structure of the plasma remains intact. Sometimes these ropes become very visible. If Earth was affecting Venus, there would be plasma tails between earth and Venus. So let's go to Kohistan.

Ancients Saw Lines of Venus

In Kohistan, there is a cave full of cave drawings. Researchers insist that a planet and star pattern depicted shows the alignments of stars as they were over 10 thousand years ago. Here is the thing that I particularly like--- there are lines drawn between Venus and Earth. They look like Plasma tails between the two planets. While this is evidence that these plasma things were around about the same time as the Carolina Bays, we have found that the plasma tails were not visible only in these ancient times.

Plasma still extends from Venus towards Earth

Today We See a Venusian Tail

In mid-1997, the Soho satellite detected a plasma structure issuing from Venus and it is long enough and in the right direction to almost reach the surface of Earth. The report described the structure as "stringy." Such a structure could only remain intact if a current were continuously flowing from Venus to the surrounding space via the plasma tail. Some researchers believe that the initiator could have been uneven electrical charges between Venus and Earth. No matter what initiated it, there is a high probability that pieces of Venus's moon were split away during the initial ionizing blast. These pieces would have fallen on Earth as a giant meteor storm. The discovery supports the idea that Venus assumed its present position in the solar system only recently, and has not yet achieved charge-equilibrium with its environment. When I say recently here I mean less than 40 thousand years.

The findings also give evidence to the probability that the reactive partner in the production of the plasma was Earth. It does another important thing as well. It makes the ancient descriptions of Venus even more believable as the planet would have looked like a huge comet in ancient times when the plasma trail was at its greatest size. It would have been a "wavy haired" planet and substantially more visible that it is today.

Venus Looked Like a Comet

Besides the wavy haired planet, various ancient names of Venus including Long Haired Star and Bearded Star, along with the other descriptors which typically symbolize comets, sound like very strange ways to describe Venus now, but that would have been exactly what would have been witnessed in the past if a plasma trail was visible at the time. In the early days, the electrical connection between Earth and Venus was possibly very pronounced and the planet was most likely closer than it is today. With these factors, that plasma tail we just discussed would have become visible and would glow just like a gigantic comet.

Venusian Heat Evidence

Today, Venus has a surface temperature of 900 - 1000 degrees F. and scientists are trying, unsuccessfully, to explain the extremely high temperatures away with a "greenhouse theory" that doesn't work. The planetary surface is so new that even the mainstream scientists are now having to devise a "global resurfacing event" [like the one presented] to explain it." We should look at all the similar myths and legends around the world describing a world-destroying catastrophe with Venus as causal agent and open up to the possibility that this well documented event could have caused havoc on Venus. Besides its lack of charge-equilibrium, Venus is totally out of "thermal balance" according to all direct observations.

There can be little doubt of the following reasons for Venus getting very hot:

- *Venus probably was closer to the Earth in the relatively recent past.*

- *Many characteristic, in-line, identically sized, blast craters point to a war on the planet. This war might have been instrumental in the eventual doom of the planet.*

- *Some huge electrical disturbance left characteristic lightning bolt marks on its surface and built plasma tails. The culprit was earth or something on the earth.*

- *Some event apparently caused Venus's moon to explode.*

- *Many of the pieces hit the Earth and Venus along both planets equatorial regions. The tight band of craters on the earth suggests that the source of the meteors was extremely close.*

- *The same event, most likely, moved Venus slightly closer to the sun. Historical references acknowledge this and the burning atmosphere enhances the possibility.*

All these things in combination initiated the thermal changes we are witnessing today. It was not the mysterious "greenhouse thing" so go ahead and use underarm spray if you want to. It's not going to destroy the earth's Ozone and push us into a cataclysmic greenhouse melt down as TAUGHT IN OUR SCHOOLS. The earth is not going to burst into flames.

Backward Spin Evidence

In order to gain some sense about its thermal problems we need to not only look at the ancient histories, or the thousands of meteoric indentions, or even the in-line blast marks on the Venusian surface. Let's look at the fact that Venus spins backwards. This would not have been primordial as the planetary spins would have taken on the general forces that were found during the solar system beginnings. Most of the planets follow this rule, but the Venusian spin was changed by some traumatic event much later. The most logical event maker would have been interaction of a close planet [possibly earth].

Venus Crater Evidence

Wait! There's more.

Besides the fact that there are very few craters on the surface of Venus which shows that the surface is very "young", we come to another curiosity. While it apparently makes no sense, almost all the craters can be found between 78 and 85 degrees in Latitude, but they can be found all the way around the planet along this central hub. Let's think about this for a minute and as we think let's look at Martian cratering.

Martian Blast Difference

If you remember from a previous section, Mars has extremely unusual cratering with almost all of the meteoritic craters located on one hemisphere. I showed that this crater grouping was not from a blast, but instead was the remains of the "old Martian surface". The other half of the planet had been ripped away. The pieces of Mars that had been blasted along with pieces of the earth that had been pulled out into outer space must be still in orbit as they would have the same centrifugal forces as the planets. They would have made a ring of planetoids and someone could call the things an asteroid belt if they wanted to.

Venusian Cratering

On Venus we have a completely different phenomenon. This ring of craters tells us that the meteors came from something orbiting very close to the planet around its equator.

I'll tell you what I think was orbiting Venus, but you have to promise not to tell anyone else, because they will think you are nuts. I think it was a moon.

Now that I've said it, I know you think I'm ready for the loony-bin, but before I go, please look at the table below. It shows the significant craters. Besides the fact that the crater density on the planet is fairly low notice that almost all of the craters are located around the equatorial region of the planet, just like I said. The ring of meteors is along the equator and varies only by +7/- 2 degrees. Moons revolve around a planets equator and if one exploded, it would make a line of craters around the equator.

Name of crater	Lat. Degrees	Long. Degrees	Diameter [KM]	Name of crater	Lat.Degrees	Long.Degrees	Diameter [KM]
Janice	87	262	10	Efimova	81	223	27
Hua Mulan	87	338	24	Tursunoy	81	229	5
Tatyana	85	212	19	Eugenia	81	105	6
Landowska	85	74	33	x	80	229	6
Ruslanova	84	17	44	X	79	270	3
Sveta	82	273	21	Nuon	79	337	7
X	82	85	12	Rudneva	78	175	30
x	82	85	2	Dashkova	78	306	45
Odilia	81	200	21	Gina	78	77	15
Lagerlof	81	285	56	Klenova	78	104	141

Weird isn't it!!

I know you have never heard about Venus having a moon and you haven't heard about the disaster that caused it to become superheated happening only 10 thousand years ago, but scientists have discovered the evidence including the stuff I previously told you, the fact that all features on the Venusian surface are "New" features, a huge split along the axis of the planet, and even the burning temperatures all point to the same thing. Twelve thousand years ago Venus had a bad time and the earth was greatly affected as well.

Venus Split Evidence

Did I say "huge Split?? Yes I did. The photograph below shows the incredible split mark across the surface of Venus.

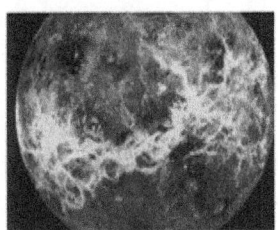
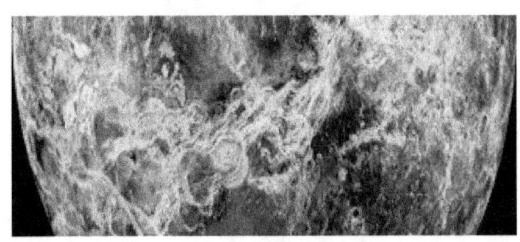

I know you are probably thinking that this picture must be a fake, because no one had EVER told you about this calamity. You are probably thinking that if there was a huge gash across the surface of Venus, it would have been headline news. Some might think that the bright area is simply a photographic anomaly, but it is not. Let me show you one of NASA' topographically highlighted detail of the area so that you will recognize that the streaks are not some photographic "ghost". The planet almost split in two from the looks of the extremely long fissures. When I say long I mean a gash that is tens of thousands of miles long.

Note the fairly thin lined fissures along the center of the photograph of Venus. [Previous right]This is not a river. It is not a lava flow. It is not some "normal" characteristic one might find on a planet. It is the undeniable splitting of the crustal surface of the planet. VENUS almost split it TWO!

Argon Evidence

To date the catastrophe on Venus, scientists use Argon. A curiosity was found by the Magellan probe. The curiosity was that the atmosphere contains high levels of the isotopes of argon, neon and noble gases. These high concentrations of noble gases could only mean that the current atmosphere of Venus is extremely young, because noble gases don't combine with other materials and escape easily into space; even with a thick atmosphere.

Venus Habitation

Here is something many researchers stay away from because of its present temperature, but evidence suggests that Venus was more inhabitable than Mars during the very ancient times. Because of an enormous amount of evidence we now can be almost certain that Venus was not always this ball of fire. In fact the flame out occurred fairly recently and many of the present surface features are extremely young. Even without its now heavy atmosphere, the climate would have been very warm during ancient times, but Venus would, most likely, have been green and beautiful. Maybe a little too warm for us to comfortably live without air-conditioning, but plants must have thrived and been in abundance. Many people lived there as well because of the relative closeness to earth and its lush landscape.

Some kind of catastrophe started to cause many volcanoes to erupt and the life of the planet began to choke. The atmosphere did not simply get thicker and thicker and finally change into an inferno as is presented in the "Greenhouse Effect Model". Whatever it was, the Venusian surface began to come alive with volcanic action from the trauma.

Venusian Landscape Evidence

Heat was noticed on both planets. Almost everything melted away on Venus, but not all evidence of Venus's previous civilization has been lost. Venus has its own, apparently man-made features similar to those found on Mars, but they are certainly not as well defined as those found on Mars simply because of the high temperatures and thick atmosphere. The picture, next left, is only one of the many scenic pictures taken by the Magellan space probe. Note the flowing river meandering down from the hills just like some beautiful spot on earth, but now the river is dried up and the vegetation is all gone. Before the atmosphere began to burn everything, the land must have been beautiful. Whoever lived near the awesome river above died along with everyone else.

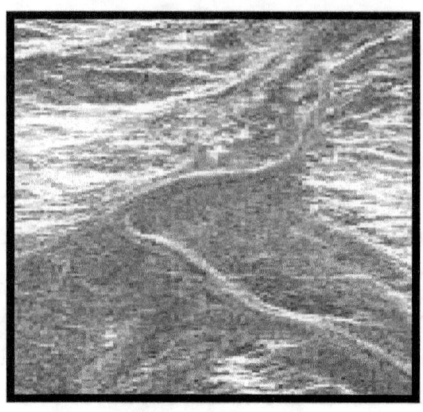 

Everywhere you look are the remains of mighty rivers that covered much of the land. Venus was truly a beautiful place at one time. It is easy to see a difference between riverbeds and lava flows or other non-water related characteristics. The meandering nature and continuously similar sized flow points only to water. Because the rivers are still recognizable, the loss of the water happened very recently. To the right is one of the dried up river deltas found. This quick dry time could not have been hundreds of thousands of years ago or the details of the riverlets would have been consumed by the planetary actions, erosion, liquefaction and other tectonic occurrences.

Atlantis and Venus

Everyone on Venus was gone and something mysterious started happening on earth. If having meteors spattering the countryside wasn't enough, the earth began to sink. All those who have heard about Plato's Atlantis descriptions raise your hands. Not you!-- you'll lose our place. I'm not going to go into this part of earth's history too much because a book could be written just on this single thing. In fact, many books already have been published containing various versions of the details of this horrible event. Anyway, Plato described an Island nation that was one of the central hubs of commerce just before the Venusian explosion time period. Other texts confirm his story and include as many as 6 other commerce centers at different locations. Over a period of years, the water level rose and rose until the inhabitants were forced to move away. The island nations went into the ocean never to rise again. Many of the survivors went to Greece while others settled in Egypt. The story sounds surreal but the timing puts it during the time of a great upheaval on the earth, so we should investigate its probability.

In order to make the story true, things that caused Atlantis to drift under the water must be found. There is a high probability that it was associated with the event that caused Venus to catch on fire. No! It was not underarm spray causing a greenhouse effect like the greenhouse fear mongers exclaim.. It was the huge explosion that almost ripped Venus in half leaving a gash that is about 25% around the surface of the planet. Whatever exploded near Venus and forced the Venusian disaster also caused the earth to be peppered with THOUSANDS of meteors along its equator and islands might have sunk.

Catastrophic Elements

For maybe a thousand years life on earth would be more unbearable than you could imaging.

- The meteorite material killed thousands

- The fires that followed killed thousands

- The sudden change in the earth's axis killed thousands

- More than likely the sun was not bright during this time because the dust particles from the meteors and plants died off. This made a terrible drought and famine which killed thousands.

- Several ancient texts indicate that 1/3 of the population was destroyed at this same time.

- Something made the ocean depths to begin increasing which killed thousands and made many more homeless.

Oh yeah! ALL of the evidence indicates that this happened about 10 thousand years ago which is about the same time as the Atlantis sinking. While everything above sounds bad the oddest is the last thing. Is there evidence that the water level did, in fact, rise, and where did the water come from? Below is some of the available evidence to consider.

Water Height Evidence

The water level increased during this time period, but it took a very long time. The water level got higher each year over a period of thousands of years. Only after a long sinking time did Atlantis submerge. One could say that its submersion was because of something we call the Wisconsin Ice Age.

To test this scientists have been testing the Atlantic Ocean. We can be fairly certain that the Atlantic Ocean is substantially higher on average than it was 10 thousand years because many have been checking and many cross referencing methods were used to test the data taken. As shown below: the average water level has increased from between 100 and 200 feet over the Atlantic Ocean's maximum height 10 thousand years. The data below is not just from one study, but is a consolidation of over 24 major studies. They all say the same thing.

Testing Method	# of studies	Ocean depth inc.
Isotope-oxygen data [volume of sediments]	5	109 feet
Calculation on the basis of gravitation anomalies	5	133 feet
Paleo-glaciological data [amount of the glaciation]	7	126 feet
Geomorphological data [ancient coastal features]	6	112 feet
Calculation on the basis of isostatic effect	1	167 feet

Prior to 10 thousand years ago, the water height fluctuated by another 200 feet as well. The graph below shows the relative water height of the Atlantic Ocean from 40 thousand years ago until today. Like the references above, the data was taken from many different studies. It shows the rising ocean described above. You will note from the table that the water level did in fact rise over a long period from about 15 thousand years ago until 9 thousand years ago. The rise is dramatic at over 400 feet during that time period, but was there an island that could have been the Atlantis of Plato?

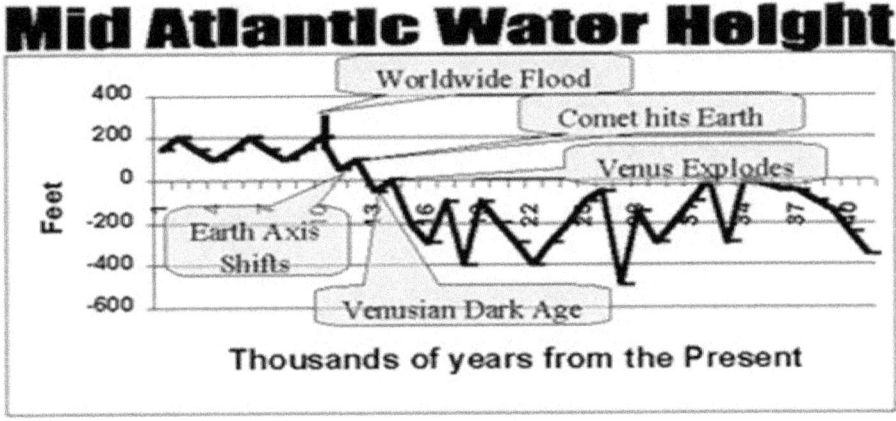

Island Evidence

The drawing below shows what the world looked like with the water level 400 feet below what it now is. A few things that pop up are that the Azores becomes a huge island in the middle of the Atlantic Ocean, the Red Sea became a river, there is a huge island in the Indian Ocean, and the Mediterranean was simply a group of huge lakes connected by rivers. Anyway, there could very well have been a huge island nation right where Plato told us it was.

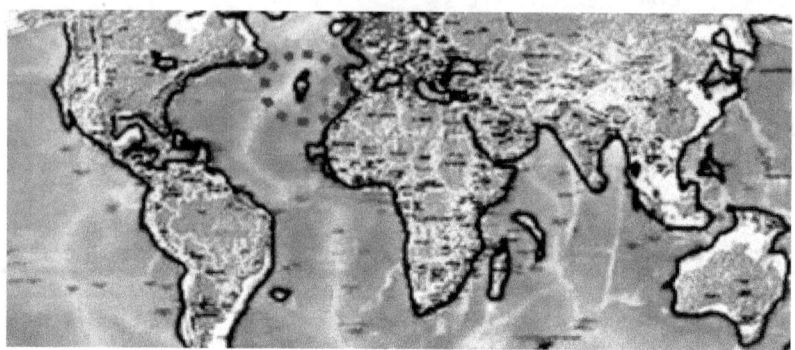

Why Did Atlantis Sink?

The Water Came From Ice

From the table below, the late Wisconsin Ice Age is timed right for the Atlantis sinking event. With the beginning and end date generally exhibiting the highest water levels and the mid-Ice Age time period exhibiting the lowest water level. The golden period of the Atlantean reign would have been 18 to 13 thousand years ago. Then the water slowly rose over the cities and the island was lost as the ice melted.

Name	Started [thousand years ago]	Ended [thousand Years ago]
Illinoian/ Great Ice Age	240	128
Early Wisconsin Ice Age	75	69
Late Wisconsin Ice Age	20	10

The following picture shows the amount of ice that was believed to have been covering the world during the last Ice Age. Much of it was converted into water to drown Atlantis. One thing to notice about this map is that there is an apparent shift in the ice edges from the current spin axis of about 30 degrees. The other thing to notice is that with all of this ice, many now covered islands would be above the water line.

I know that the apparent equatorial line is not that indicated for the Carolina Bays so don't give me a hard time. From the table above, the Ice age started about 20 thousand years ago. It is reasonable to believe that at that time the earth axis was as shown below and moved into the position indicated by the Carolina bays by the time of the Great Venusian Disaster. Also you might have noticed that the water level began to get higher about 15 thousand years ago which was before the Venus thing.

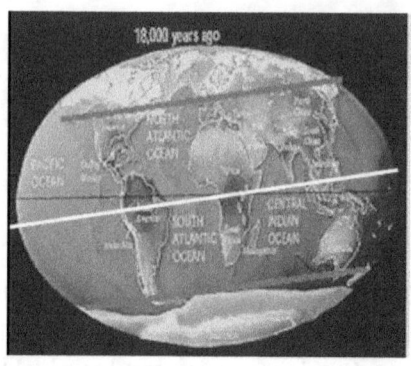

Let's examine this Ice Age thing for a minute. Although I do believe that the earth was colder some of the ancient time period, the whole glacier thing should be looked at a little more closely.

The Glacier Error

Have you wondered how the Glaciers could have moved thousands of miles <u>uphill</u> as commonly attested to by "glacier experts"?

The well-known investigator, Herbie Brennan, proposed that water going uphill just has to be wrong. Just think about the supposed onslaught of Glaciers that covered much of Europe and the United States during portions of our history. Many times the only way for the glaciers to have gone where they did was this idea that glacier water could go uphill many hundreds of feet. Physics has a hard time with the concept and so should you.

The reason people think glaciers did this impossible task is that rocks were left on top of mountains and in valleys that have been scraped before being deposited in unusual places. These rocks are called erratics by the glacier scientists. Unfortunately these same erratics have been found in the Sahara Desert, in Uruguay, and in Mongolia even though no one believes that Glaciers passed over those areas. For someone to present massive glacier movements covering the northern part of our planet as a fact, seems inappropriate. Don't through one science away to support another science. They need to work together. As part of the glacier thing, Mammoths were depicted lumbering through thick glacier fields. [No Way!!]

The Mammoth Error

Have you ever wondered why they always show Mammoths in snowy regions? Certainly they had 20 inch long fur just like the long fur associated with orangutans and the huge manes of lions, but that doesn't mean that the animals lived in snow. Elephants of all types need huge quantities of vegetation to survive. This huge version must have had an insatiable appetite for flowers and such. Whenever remains of foods are found in mammoth carcasses, the foods are flowers and vegetation. They could not have lived in the cold regions of the world. Mammoth bodies have been found in Alaska and Siberia, but, "guess what?" those places must have been warm if mammoths were there. For someone to tell you otherwise is inappropriate.

Axis Movement Error

Let's first start out with a very important statement.

THE GLASIERS GENERALLY DID NOT MOVE.

If the glaciers didn't move, you might be wondering just how the ice got there. The answer might be found in the earth's axis. I don't mean the shift in the crustal areas with respect to the earth's spin axis, I mean the spin axis of the earth with respect to the sun. As shown below, today our planet has a 23 degree tilt with respect to our revolution axis around the sun. The areas with the most variation with respect to amount of daylight are located at the poles. Sometimes these areas have 6 months of darkness and with the darkness comes bitter cold. A spin axis perpendicular to our revolution would make the mildest pole temperatures while more incline means more cold.

The 23 degrees is far from stable and has shifted many times. If the shift increases the incline we have an Ice Age. The shift that increased the water available and destroyed many island nations would have been a shift towards a decrease incline. This is much more in line with physics than the water going uphill thing and we can date just when the earth's axis became less declined and has generally continued its 23 degree angle.

Another Possibility

The other possibility, I mentioned before. The earth spin axis with respect to the sun might not have changed, but instead, the earth's orbit could have been farther away up until about 10 thousand years ago. Whether the shift was caused from the interaction between Venus and Earth or the earth shift came first and began the whole earth Venus thing, is not known, but any of these things are more probable than glaciers moving uphill as you have been taught.

City Evidence

There is evidence that this first sinking was slow and did not cause the massive tidal waves and upheaval felt during the "worldwide flood" that we will study later in this book. The relative slowness of the rising waters allowed some of the artifacts from before this flood to remain. Some of the remains of the magnificent preflood cities can still be found today, 12 thousand years later. We don't find one city, we find a dozen. Unfortunately you have to look under the water. One of these many undersea cities could have been the famed Atlantis. Surely that wasn't its real name, but Plato burned its name into our consciousness, so I will use that name to generalize the island nations of our past. Below are some city artifacts that have been found underwater. The evidence suggests that the majority of the destructions occurred about 12 thousand years ago. Many cities, evidently, ended up underwater during this "special" time and the water never receded down to its preflood levels again. The fact that these cities were already submerged actually saved their remains while those cities above the water were almost completely destroyed in the huge tidal wave actions of the more famous worldwide flood to come.

City near Morocco-Roads and buildings were found in 60 feet of water off the Moroccan coast.

*North Sea City-*In 1954 the remains of a city was found in 50 feet of water. Molded slabs of firestones for road pavement and smelted ore products were found. The estimated age of the city was at least 3500 years old, but it could have been much older. **[I say, 8 thousand years older.]**

City off Coast of Spain-   Off the coast of Spain, the remains of a city that sank in the ocean thousands of years ago was found in 1973. According to "United Press International", the remains included walls similar to that shown next left, large columns, and roads.

Off the Japanese Coast-   Huge   underwater   "temple-like" structures have been found. Possibly they are the remains of a thriving city that flourished thousands of years ago. A huge temple structure and what appears to be a roadway are being investigated. The photo to the right shows some of the regular "Steps".

*Portugal-*Four thousand miles off the coast, Russian scientists found buildings made of strong concrete and even plastics. Along with the buildings, they found the remains of city streets. They brought up a statue from the area for study.

*Black Sea Find-*Off the coast of Turkey, large blocks associated with a city that sank thousands of years ago have been found.

*Off the Coast of France-*A flight of stairs was discovered in 1964 leading downward off the coast while researchers were in the bathescope Archimedes.

*Greece-*While trying to find the arms of the statue Venus, Jim Thorne found an underwater citadel near Melos. The remains were several hundred feet deep with roads going even deeper.

*Malta Waters-*Cart tracks going everywhere on the island also go straight into the Mediterranean Sea. The roads appear to connect the island with the mainland. This "center of the Mediterranean"

Island was probably once part of the mainland and was part of a much larger civilization before the initial water rise.

Rhode Island-In 1958, near the Brenton Reef, skin divers reported finding a conical shaped tower estimated to be 50 feet high in 90 feet of water. Each of the stones on the tower was estimated to be the size of an average desk.

Peru-Photographs of carved columns were reportedly taken at a depth of 1000 fathoms off the coast of Peru. Sonar soundings also indicated possibility of buildings at the site.

Bahamas- A pyramid, roads, columns, a dome, rectangular buildings, a statue, unidentified metal objects, and a 1600 ft. long underwater wall made of limestone blocks which are 15 to 20 foot square have all been found off shore. The picture below shows a section of the wall.

Plato's Atlantis

Ok! There are many underwater cities, but we need to go back and first get a feel for what Atlantis was all about before we continue because we have heard about Atlantis as a myth so long that it sounds funny talking about such a place. I know this seems like a bizarre subject, but you have already gone through many bizarre subjects and you are still reading this history, so who's the oddball?.

Don't get me wrong. I haven't been trying to find strange elements of conjecture to push into some type of fairytale. I am bringing you a cross comparative collection of elements to support a probable history that just happens to link some of the stranger writings and physical evidence together with more mainstream science and religion. Atlantis fits this mold.

Everyone has heard about Plato's Atlantis description covered in the book "Timaeus", but there is one aspect of his history that should be brought out a little more. It deals with the dual flood concept which is important to understanding Earth's development.

Timaeus 22d-23c

*And the reason is this. There have been and will be many different calamities to destroy mankind, the greatest of them by fire and water, and lesser ones by countless other means. But in our temples we **have preserved from earliest times a written record of any great or splendid achievement** or notable event which has come to our ears whether it occurred in your part of the world or here or anywhere else; whereas with you and others, writing and the other necessities of civilization have only just been developed when the **periodic scourge of the deluge descends, and spares none but the unlettered and uncultured**, so that you have to begin again like children, in complete ignorance of what happened in our part of the world or in yours in early times... [**This indicates that**

141

the "Atlantis sinking" was not the last major catastrophe encountered by the Egyptians and Greeks.]

*You remember only one deluge, though there have been many, and you do not know that the finest and **best race of men that ever existed lived in your country**; you and your fellow citizens are descended from the few survivors that remained, but you know nothing about it because so many succeeding generations left no record in writing".* **[The most obvious reason that the Greeks had no direct knowledge of the Atlantean refugees colonizing Greece would have been that another terrible flood or cataclysm occurred a long time after the Atlantis sinking. We have all heard of this second flood as it essentially covered the entire world a mere 3 to 4 thousand years after the Atlanteans colonized Greece.]**

Plato's Flood Timing

Even Plato's timing matches the evidence and the presented timeline. I know you have been told that Plato's Atlantis sank 9 thousand years ago, but that is not what his history stated. Assuming the information was obtained about 1000BC and the Egyptians were instructed for 8000 years, and Greece was instructed 1000 years before that, then the time of the sinking was between 12 and 13 thousand years ago, which is the time period we are investigating. Let's read what was said exactly in both "Timaeus" and "Critias".

Timaeus. 23d-24a

*"Solon was astonished at what he heard and eagerly begged the priests to describe to him in detail the doings of these citizens of the past. "I will gladly do so, Solon," replied the priest, "both for your sake and your city's, but chiefly in gratitude to the Goddess to whom it has fallen to bring up and educate both your country and ours - yours first, when she took over your seed from **Earth and Hephaestus**, ours **a thousand years later**. The age of our institutions is given in our sacred records as **eight thousand years**, and the citizens whose laws and whose finest achievement I will now briefly describe to you therefore lived **nine thousand years***

ago; we will go through their history in detail later on at leisure, when we can consult the records."

Critias

"Let me begin by observing, first of all, that nine thousand was the sum of years which had passed since the war said to have taken place between those who dwell outside the Pillars of Hercules and those who dwell within." [With Plato living almost 3 thousand years ago, 9 thousand years from Plato was about 12 thousand years ago.]

Plato Wrote about a 2nd Flood

Not only did Plato talk about the flood that destroyed the Atlantean Island, he also told about the second more serious one. Let's continue reading.

Timaeus 25c-d

*"At a later time there were **earthquakes** and **floods** of extraordinary violence, and in a single dreadful day and night all your fighting men were swallowed up by the earth, and the island of Atlantis was similarly swallowed up by the sea and vanished..."* [This was not referring to the slow increase of flood waters around Atlantis, but a separate and violent flood from which Noah survived according to Biblical accounts.]

Critias

"Atlantis once had a greater extent than Libya and Asia and afterwards sunk by an earthquake." [The description of sinking by earthquake rather than simple flood seems to go along with other data.]

"Many great deluges have taken place during the nine thousand years." [Of course one worse than the Atlantean flood occurred later and practically everyone was drowned. We will find that this terrible flood occurred about 9 thousand years ago or about 3 thousand years after Atlantis disappeared into the sea.]

Ice Age Proof

One reason that the water level never came down to its pre "Atlantean flood" levels was "less ICE". If you recall there was something we call an ICE AGE about this time. I stated that Venus probably moved in closer to the sun around this time and that earth was affecting the planet during this time. The evidence suggests that earth was possibly pushed in closer to the sun a little just as Venus was being pushed toward the sun. The effect was obvious. As earth went closer, some of the Ice turned to water on the earth and stayed water until today.

Evidence of a Shift

One thing that could cause a change in water height is a change in the earth's rotational axis. From several sources, we can gain a good picture about the earth shifting on its axis during the Venusian adventure. The texts tell us a gruesome tale about 1/3 of all life being killed during this struggle for earth to settle down. .

Enochian Evidence

In the book of Enoch, the event is described vividly along with the cause of the event.

According to "Enoch", men had learned too much from the angels, and would not repent of their evil ways. A third of the animals and people all died due to this mistake.

*Enoch 67:14-"The water of the springs shall **again** undergo a change, and become frozen." [**This ancient writing tells of the drastic shift in the climate that may have come from the rotational axis shift. It also indicates that this type of shift had happened before. I know it also sounds like the cyclic Ice Ages of the Earth, but what causes those things?**]*

*Enoch 64:1-[**This ancient Jewish text confirms the event**] "In those days Noah saw that the **earth became inclined** and that destruction approached. -The earth labors and is violently shaken. Surely I shall perish with it. -There was a great perturbation on the earth. [**The incline change denoted a shift and the perturbation denoted gigantic earthquakes. This verse was before the worldwide flood time.**]*

Other Written Texts

We find the same story from Greek and Egyptian historians.

*Herodotus-According to ancient Egyptian historians and written down by the Greek historian Herodotus, "The sun did not always rising in the same place." [**The only way for that to happen is if the Earth axis shifts as would be evident from a crustal jump.**]*

*Egypt-The Harris Papyrus states, "A cosmic cataclysm of fire and water followed with the **south becoming the north** and the earth turning around." [**This text dates the shift as being after the cosmic cataclysm of Venus**]*

Physical Evidence of the Shift

Before you disregard the notion of the earth axis shifting from a multitude of meteors hitting it, here are a few of the many indications of the event.

Straight Line Distribution Evidence

I told you to remember the Carolina Bay direction presented earlier. Well, here is why it should have been noticed and remembered. If we look at the distribution pattern of the "Carolina Bays" it becomes apparent that there is a straight line of events that occurred around the world. The direction of the impact axis is shown below left. In fact, if we follow the line around to the other side of the world, we find more evidence of massive meteorite strikes that occurred at the same time in **Australia**. You say SO WHAT!

Besides giving us knowledge of a tremendous meteoritic event, the Carolina Bays provide us evidence of our last rotational shift on Earth and they even give us a good approximation of the previous axis of rotation for the Earth. That is because the density of the bays and the evidence in Australia show a "straight-line" distribution pattern that is consistent with a bombardment along the equatorial boundary. If we consider the impact density line as the "Old" equator, a shift of about 30 degrees in the rotational axis has occurred since the bombardment as shown in the picture to the right. Above, the globe has been separated at this ancient equator. Note that along the polar path there was not much land. Also note that eastern Alaska and Siberia are well away from the Arctic Circle, which allowed huge herds of Mammoths to dine on flowers in those areas just before the shift. The shift froze them solid

As we look at the shift across its equatorial cross-over point, we can quickly see that the Peruvian [PreInca] and Afghanistan [Aryan] societies would have been thriving prior to the shift. They were both situated on the tropic of Cancer and Capricorn. After the shift they would have had to migrate.

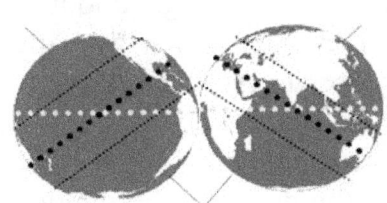 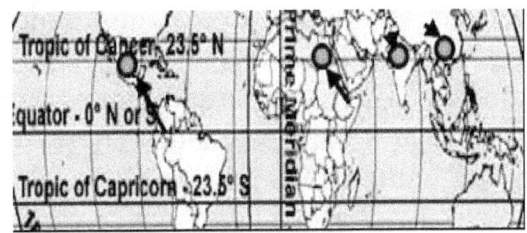

Migration Evidence

After the shift, the last shift the Afghanistan region moved north away from the comfort of the Tropic of Cancer and the garden spot moved into India as did a large segment of the world society. Peru moved into the Equator, so the Peruvian inhabitants moved north to Central America. The map above shows the general locations of major civilizations after the shift. The migrations shown all occurred about 10 thousand years ago.

Two Flood Evidence

While I'm talking about a great flood that overtook much of the lowlands of the earth, there is strong evidence that there were actually two major floodings that occurred within about 2 to 4 thousand years from one another. This first one was important in that many survivors were able to escape to the Americas, Europe and Egypt. Even though many did survive the first one, almost all the inhabitants were destroyed in the worldwide flood that followed fairly soon [3 thousand years] afterwards so don't get the two confused.

More Timing Evidence

I know I've already stated that the particular flood we are discussing here occurred 12 thousand years ago, but I thought it would be nice to read some of the historical insights of the flood and the island nation.

Plato made the city of Atlantis famous with his treatise about the huge civilization beyond the Pillars of Hercules which he called Atlantis. There are actually many references to Atlantis in ancient tests besides that of Plato. These references can be found from Egypt, India, Greece, China, Peru, Ethiopia, Persia and other places around the world. The evidence confirms the date and the incident. Here are some of the testimonies that let us know "when" Atlantis sank.

Greek Timing Evidence

Two historians from Greece tell us the same thing. The historians were Plato and Strabo.

*Plato-From his third hand knowledge, Plato indicated that 9000 years before the time of Solon, or about **12 thousand years ago**, the city of Atlantis sank*

*Strabo-This Greek historian wrote that about **10,000BC** the Greeks fought the Tartessians [Atlanteans] beyond the Pillars of Hercules which must have been before its' sinking.*

Egyptian Timing Evidence

Evidence can be found in Egypt. Here are some excerpts from Manetho's works and several extant papyruses [or is it papyri?].

*Manetho-This Egyptian historian stated those 5800 years before, Menes, the first Egyptian king, was from the dynasty of "The Spirits of the Dead". The time would be about **11,000 BC**. [The dead probably is reference to the death of Atlantis.]*

More Egyptian records-A papyrus in the museum of St. Petersburg stated "Land of Atlantis, whence had come the ancestors of the

Egyptians 3350 years ago---the sages of Atlantis flourished during a period of 13,900 years" [Although the time of the Atlantis destruction cannot be determined from the 3350 year number, this does tell us that if the submersion was 12 thousand years ago, then Atlantis became powerful about 24 thousand years ago.]

Atlantis Time-line-*In the "Manetone Papyrus" which can be found in the British museum, the Egyptian historian named Manetone provided some insight into the lineage of the kings of Atlantis. He indicated that the Atlantean kingship went back for 13,900 years before the initial beginning of the Egyptian kingship. King Thoth was the first king from Atlantis according to his own account in the "Emerald Texts". If we place Thoth at the worldwide Flood it would place the Atlantean kingship starting about 20 thousand years ago.*

"Sent Papyrus" Evidence-*According to the "Sent Papyrus", which is the oldest papyrus known to exist, and is resident in Petersburg Museum, in Russia, Pharaoh Sent dispatched an envoy to search for Atlantis from where the Egyptian forefathers had supposedly come to Egypt some 3500 years earlier bringing wisdom.*

Tibetan Timing Evidence

*From Tibet, the "Book of Dzyan" says, "Large areas of land sank in **9564BC**."* [I'm not sure why their date is so precise, but that would be 11,567 years ago.]

American Timing Evidence

This evidence can be found from the Maya, the Aztec and the Caribbean.

Codex Troano-*This Mayan text indicated that the destruction of **Azatlan**, by sinking, occurred 8,060 years before the writing or a little over **11,000 years ago**. It also indicated that millions of people died in the cataclysm.*

Aztec History-*According to the Vatico Latin Codex the flood that sank **Azatlan** occurred **13,205 years ago**.*

United States-*Off shore excavations near the Atlantis site off the coast of the United States has dredged up calcareous disks about*

*15 cm in diameter and 4 cm thick with a round hole in the center. Radio carbon dating indicates that they were formed on dry land over **10 thousand years ago,** which provides still another level of confirmation of the catastrophe date.*

Pleistocene Timing Evidence

Geologic evidence concurs that much of Earth was destroyed by fire, volcanoes and ice about **10,000 BC**. This was the ending of the *Pleistocene Age.* Volcanic eruptions spewed everywhere and blocked the sun. This may have added to the coldness and the Ice areas, but it would have been too short lived to support low water levels for thousands of years.

Seer Timing Evidence

*Edger Cayce had a vision about Atlantis. -His psychic readings indicated that before **10,500 BC** flood consumed Atlantis.* [OK! He could have read Plato's work, but this guy had many accurate visions of the future that have come true, so don't discount data just because it may at first seem dubious. Certainly recognize the source and temper its importance, but don't ignore date you don't like.]

Water Temperature Evidence

We don't have to simply rely on –

- *Dozens of water height studies,*
- *Dozens of ancient historical references,*
- *Physical evidence associated with the Carolina Bays,*
- *Suspect migration patterns,*

That certainly is not enough to make us think that the earth shifted, but what if we could show that the water got hotter all of a sudden and stayed hot. We can be fairly certain that the water temperature in the Atlantic Ocean abruptly increased as the Earth's axis shifted. The National Geophysical Data center provides the following information concerning the water temperature over the last 100 thousand years.

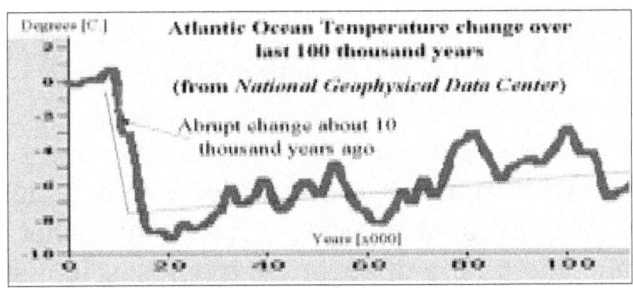

Note the change that occurred over about a 5 thousand year span from 15 thousand years ago until 10 thousand years ago. The average water temperature increased by 10 degrees Centigrade or 18 degrees Fahrenheit. One reasonable way for that change to occur is for the earth axis of spin to change bringing more of the Atlantic Ocean into the equatorial region as discussed above. After the Venusian catastrophe and the changes that occurred on earth, there was a short respite. People began to get confident again and life became normal. For a thousand years life was ok again, but the world would see another drastic change a mere 9 thousand years ago. This time the entire world flooded.

8000BC Worldwide Flood

Some scoff at the idea of a worldwide flood destroying everything, but there is a huge amount of evidence. The thing that is revealing in the evidence, but totally disregarded by most scientists is that the flood occurred during a time when the earth had flipped on its axis. Here are some of the things we know about this important time in earth's development.

A worldwide flood occurred 10 thousand years ago at the end of the Pleistocene Age

At that same time there was such violence in the atmosphere that hundreds of animals were twisted together in large piles of death

At this same time the western coastline of South America rose out of the ocean.

At this same time there were surviving humans around the world

At this time there were surviving animals trapped on isolated regions which still show graphical differentiation

There may have been several things going on at this time, but flash freezing of Mammoths and huge piles of different animals twisted together with sea creatures and land animals mixed together occurring at exactly the same time should tell us something. Localized atmosphere changed quickly and polar ice quickly melted and refroze somewhere else. While all this was going on and tremendous change in weather patterns along with unbelievable tsunami absolutely flooded the inhabited portions of the world about 9 thousand years ago.

Noah

One cannot discuss this earth changing event without talking about Noah and the arc, but we need to look at this event with reason or the truth will stay hidden and science and religion will continue to be separated by a gulf of interpretation. While Noah did survive the flood, around the world we find out about the other "Noah-like characters". The various Jewish texts indicate that God selected a few of the Adamic men [**those who were direct descendants of Adam**], had them gather up various animals and he flooded the world. The Sumerian text says that the survivors carried the seeds of all animals. This "seed thing" sounds like the genetic codes of the animals were all that came on the arc in the Sumerian version, but the Biblical version specifically indicates quantities of animals to take on board. The real thing that should be remembered is that <u>some people and some animals</u> survived. While the above seems obvious, the other thing that can be considered as truth is that Noah's family members were survivors.

Noah Was the Only Survivor

As a contradiction to the concept of people around the world surviving the flood, we are told that Noah was the only survivor, and so he was. If we read the fine print we can see that Noah was the only "Adamic human" to survive. Then the Bible specifically indicates that only people on dry land died. You should know that there were very few pureblooded Adamic humans on the earth in the first place and only 8 of Noah's family survived. That is not to say that there were no survivors. Around the world tiny groups of people survived and replenished the earth. The others that survived were regular people, but Adam and Eve were not their sole ancestors. They were mixed with the 6th day humans the Bible told us about. I'm not getting into this concept in this book, but one should recognize that the Biblical history and scientific testimony should show the identical elements. If we assume that the flood did occur can we say that Noah and his family were the only survivors?

I say No, but that does not mean that the Bible is in error.

Before some of you start saying I'm indicating that the Biblical history is a lie, just hold on just a little bit longer. The stories that will be presented are from around the world. There must have been additional surviving groups and each had the mission of carrying animals with them. Specific animals were deposited at locations around the world that would become their homes and the various people were carried to different parts of the world and there they established homes. As with the Noah story, the survivors probably did not venture far from their pre-flood homelands when the floodwaters finally subsided.

Is the Bible Mistaken?

The Bible indicates that all humans that were on dry land perished in the flood, but it stops short of indicating that no other humans survived. Here is, I believe, the most reasonable hypothesis. <u>Noah and his family were the only pure humans that survived</u>; therefore, all the other survivors were **hybrid** humans, so an indication of Noah and his family being the only "Chosen ones" or pure Adamics is both reasonable and correct.

Genesis 7:21-22 And all flesh died that moved upon the earth, -- every man--- that was in the dry land, died. [All flesh died that was on the dry land, but survivors on boats that could have possibly withstood the onslaught of massive tidal waves and may have survived. If it were not so, the Bible would not have identified only the humans on dry land as those being killed.]

We will look at some more of the proof to this hypothesis as we go along. Some of the proof is provided in the flood stories themselves. As you read them notice references to at least 2 different types of humans. One type would be Adamic humans and the other group would, therefore, not be pure Adamics. Almost everyone that survived was not a pure Adamic human. If it were not so, why would God have a chosen people?

The texts from Genesis show details of the initiation and destruction of the flood. One verse has been cause for some to question the authority of the Biblical history concerning the flood. One interpretation of Genesis 7 indicates that all mountains were covered when the water rose 15 cubits. Of course that would not be correct nor would any early writers believe that so don't throw out the truth when confusion arises. Let's go back to the Hebrew.

Genesis 7:20-23 "**Fifteen cubits**" *upward did the waters prevail; and the mountains were covered.* [Although many use this verse as a mistake, the Hebrew actually says "ASAR cubits" or "multiples of ten cubits" rather than "15" so it really says the water rose to whatever height it took to cover the mountains.]

Whenever something sounds wrong, don't throw away the remaining information, try to see what is wrong and try to get the "Good" out of the data.

With all that being said, let's look at the stories themselves and know that the water was much deeper than 15 cubits and just about everyone and everything drowned.

Real Live Animals on the Arc

There are some who say that God can do anything and it doesn't matter if the cause or the effect doesn't make sense in the scientific world. Then they say, "Don't try to change what is written in the Biblical history." Those same people indicate that it would have been possible to push into Noah's 4-story high, 400 foot long vessel, all of the types of animals because the Bible indicates that at least two of every animal type entered the arc. That would be over 13 million animals and insects. Even if you threw out the insects there would be over 3 million species of animals mooing and kicking in the less than spacious boat. [That would be less than a half square inch per animal and man, including the elephants. Also in this space the food needed for a year long journey had to be squeezed. By the way I'm not saying Noah could not have done this, I just think we should add in some reason.]

*Genesis 7:2-4 You must take with you 7 of every kind of <u>clean</u> animal, 2 of every kind of **abominable** animal, the male and its mate, and also 7 of every kind of bird in the sky, male and female, to preserve their offspring on the face of the earth. For in seven days I will cause it to rain.*

Some Agree With ½" Space

The reason for the insistence that ALL animals were on the Arc is that if the animals were not on Noah's ark, and only things on the arc survived, then the animals could not have survived the flood? To bring every bird, land animal, and insect, seemingly, would have been impossible. If we add to this statement we should say that if God was going to do the impossible, he would not have forced Noah to do all the collecting and would have simply zapped them into existence after the flood. I think it is pretty obvious that something else was going on here because different animals types are only found in specific places in the world today. America has no Platypuses and Australia has no African elephants.

A part of the answer is that there were more ships being built around the world, but that still doesn't explain the vast quantities of animal types to have survived.

Fewer Animals Probability

If we give credence to the previous statement about multiple boats, the situation is helped, but it is still seemingly impossible because of the duplicate animal types that are found around the world. There is another even more responsible answer, but it will take some explaining.

There were probably not very many animal types remaining on the earth after the Venus incident.

Well before the worldwide flood, during the time of the predecessor flood, ancient texts told us that 1/3 of everything was gone. One might assume that animals were included as part of the previous destruction. It would not be impractical to believe that more than 1/3 of the animal types met extinction because of other similar destruction periods. With only a few animal types to worry about, it would not have been impossible to carry them on board and satisfy the ancient texts. This theory, of course, has a different problem on the outcome side, but first let's look at one of the ancient verses that tells us about a minor extinction period before the flood.

*Jasher 2:5-6- "-And the Lord caused the waters of the river Gihon to overwhelm them, and he destroyed and consumed them, **and he destroyed the third part of the Earth,** and notwithstanding this, the sons of men did not turn from their evil ways, "* [This isn't talking about the really big flood. It's talking about the flood before the flood.]

Where Did NEW Animals Come From?

If only a few animals were in the world at the time of the flood, then where did the "New Animals" come from? The answer, most likely, is that, just like before the flood, genetic manipulation was rampant after the flood. Animals must have been created or modified one after another in order to re-produce the 1 million animals, 5 million insects, and 2 million plant species we have today. The microscopic animals were probably modified as well, so you can see the task was a big one. Even if we assume that there were 100 surviving boats and each boat carried 10 thousand animal types there wasn't enough room. From the distribution of animals around the world it is evident that many of the animals on-board each of the vessels were the same type, so we might be able to account for about 1/4 million animal types fully rescued [not including insects]. The remaining 3/4ths of the species would be left to be re-engineered after the flood.

I know no one wants to believe this possibility, but; if one believes the Biblical account to be correct; if widely variant animal types are found in isolated regions around the world; if there are over a million animal species in existence; and if God didn't simply zap the creatures here, the regeneration theory seems to be the most logical story. An entire book could be written about the flood its aftermath, but we will move on, because that is only a small time in earth's history. Before we do, there is a reasonable question to ask That is, "Did the worldwide flood actually did happen?"

Did This 2nd Flood Happen?

Maybe we shouldn't just assume that this character named Noah piled a bunch of animals in a boat to float for a year. There should be evidence. Thanks to the efforts of a historian by the name of Aaron Smith we have a count of flood stories. At least those he could find.

His count came to 80,000 flood stories in 72 languages

Practically every culture, every religion, and every history contains stories about the "third" creation of man and his eventual destruction by a worldwide flood. I have selected and, very briefly, paraphrased thirty that showed strong similarities.

Note the similar reasons for the flood. Note the identical use of birds to test the land. Note the similar discussion of gods controlling the land before the flood. Note the concept that this was the 3rd period or third creation that was destroyed. Note the concept of **two types of humans** *being saved.*

There can be little doubt that these are the same story. If the flood story is so similar around the world is there reasonable doubt of the flood itself? I think the answer is No.

Middle Eastern Flood Stories

Jewish Flood [from various ancient texts]-*The rulers before the flood were worshiped as* **gods**. *These rulers were* **giants.** *The flood occurred because the* **third race of men** *was disobedient to God. It rained for* **40 days**. *Noah saved 7 of every "normal" animal and* **2 of every "genetically engineered" animal**. *He also saved his immediate family. Afterwards he* **sent birds** *to test the land and God sent a* **rainbow** *as a sign. Some* **Anakim gods survived** *the flood as well. Later, man was punished again and many became like monkeys when the Tower of Babel came down.* **[Now let's compare everyone else's version.]**

Chaldean Flood-*There was a worldwide flood. Survivors of the worldwide flood sent out* **birds** *to see when the land appeared. The* **gods** *also survived.*

Sumerian Flood-*The* **gods** *controlled the world before the flood. An arc was built and the seed of every animal was transported along with the human survivors.* **Birds** *were sent out by those saved from a worldwide flood. God sent a* **rainbow** *afterward. The* **Annunaki gods also survived**. *[Pretty close to Anakim isn't it?]*

Hindu Flood-Three *worlds were flooded. [Three creations of man]Many were saved each time.*

Bay of Bengal Flood-*According to the India Indians, men grew* **disobedient**. *Puluga, the creator, sent a flood to destroy everything. It covered the whole land. Only two men and two women survived. That was the last time God and man spoke face to face.*

European Flood

Greek Flood-*Greek* **gods** *ruled the land before the great flood. The flood occurred during the* **3rd Age** *of man. There were 9 survivors. The* **gods survived** *as well.*

Celtic Flood-Giants *controlled the world that ended by flood. A worldwide flood came. 2 men survived the world destruction.*

Lithuanian Flood-*God was upset with men because of sin and continuous war. The world was controlled by* **giants**. *A flood destroyed the world. Afterward, God sent the* **rainbow**

Harappa-In ancient Pakistan, the Flood story was depicted as this boat full of animals. [See below left]

 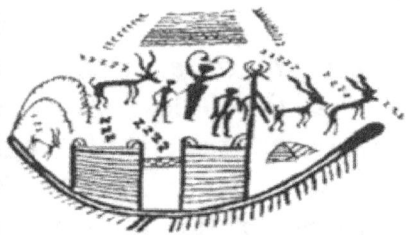

African Flood Stories

Egyptian Flood-*Pyramid texts states –the **3rd Period** was called the **Golden Age** of man and it was destroyed by a worldwide flood. In ancient Egypt, the Flood story was depicted as a boat full of animals as shown above right.*

Southwest Tanzanian Flood-*The world began flooding, God told 2 men to go into a ship and take with them all sorts of seed and animals. The flood covered the mountains. After a while, the men sent out a **dove** to see if the water had dried, but it came back to the ship, then they sent out a **hawk** that didn't return and they knew the land was dry, so they disembarked.* [Similar to the raven dove Jewish version]

Lower Congo Flood-*The sun met the moon and threw mud, making it dimmer. While the moon was dim, a huge flood occurred. Men put their milk sticks behind them and were turned into **monkeys.** Later a new race of men was created.* [We will see later that the milk stick thing is important with respect to the Tower of Babel, but the idea that the moon was dim before the flood seems to show knowledge of a Dark Age preceding the event.]

Victorian Flood-*The Creator, named Bunjil, was **angered because of the evil** that people were doing, He caused the ocean to flood by urinating into it. All the people were destroyed except those whom Bunjil loved. These people finally became stars in the sky. **[Adamic humans had the ability to go to heaven]** Besides this group a man and woman also survived who climbed a tree on a mountaintop. **[Non-Adamic humans did not have the ability to go to heaven]** From those two, the present human race descended.* [In

160

this story there are two types of humans that survived the flood. The urinating part is slightly different.]

Far Eastern & Pacifica Flood

Chinese Flood -*In the Chinese flood story, there was a god named "Gong Gong". He was ordered by the head of the gods to create a flood. It was as a* **punishment for human misbehavior**. *The flood lasted 22 years, until the hero started to dam the waters. The hero was killed for this act, but from his corpse sprang a son who finished his father's project.*

Burmese Flood-*There was a worldwide flood. Nine people were saved. They sent out* **birds** *to test the land.*

Tibetan Flood -*Even Tibet was inundated with water. God, Gya, had compassion on the survivors and drew off the waters through Bengal. He sent teachers to civilize the people again. After the flood they were little better than* **monkeys**. **[Again the reference to monkeys will be clearer as we discuss the Tower of Babel, but in this story, the idea that gods continued to teach humans after the flood is evident.]**

Philippines Flood-*Two humans were saved from a worldwide flood.*

Maori [New Zealand] Flood-*Some people were saved from a worldwide flood.*

South & Central American Flood

Totonac –Mexican Flood-*A man was warned of upcoming flood and built a boat. He was saved from the worldwide flood. After the flood, he sent out a* **buzzard**_*to test the land. The boat finally rested and God reversed man's face and hind parts and turned him into a* **monkey**.

Aztec and Mayan Flood -*The* **3ʳᵈ Period** *was called the* **Golden Age.** *It was destroyed by storms.* **Nene** *[Aztec Noah] was saved. He sent* **birds** *out to test the land.*

Yamana Flood-The moon-woman Hanuxa caused the flood because she was full of **hatred against the people**, *especially the men, who had taken over the women's secret kina ceremony and made it their own, so she let it snow so much that ice came to cover the entire earth. When it melted, it rapidly flooded all the earth except for 5 mountain tops. A few people survived on five mountain tops. [Very similar to Biblical and geological evidence. People became evil; the ice age came; the earth apparently shifted or got warm all of a sudden; and the flood killed all but a few.]*

Andaman Flood-The Supreme Being, Puluga, lived in the sky. Puluga created the whole world and man. **Man began to forget his creator**. *Puluga became annoyed and sent a flood that covered the whole earth and wiped out the race. Four people escaped.*

North American Flood Stories

Jicarilla (Apache) Flood *-Before the Apaches emerged from the underworld, there were other people on the earth [a previous creation of humans]. Dios [God] told an old man and old woman that it would rain* **forty days** *and nights. [The old man and woman were like Noah and his wife-Adamic humans] Other people were warned to go to the tops of four mountains and not to look at either the flood or the sky [These would* **not** *have been the Adamic humans]. The people didn't believe the old couple. When the rains came, only a few people made it to the mountain tops and shut their eyes. Those who looked at the flood turned into fish or frogs; if they looked at the sky, they turned into birds. After eighty days, Dios told the 24 people remaining to open their eyes and come down. These 24 people went into mountains.* **Eight other people** *[the Adamic humans] survived the flood that were able to travel because they could look to see where they wanted to go. These people told the Apaches about the flood before going into two mountains themselves. Around the turn of the millennium, the surface of the earth will again be destroyed, this time by fire.*

[This story is special in that there were two types of humans that survived the flood. There were only 8 who could see where they were going just like the 8 Adamics that survived in the Biblical history. The other 24 non-Adamics had to keep their eyes shut and trust that God would help them.]

Blackfoot Indians   Flood- *Some were saved from 2 separate devastating floods. The* **thunderbird** *was instrumental in their rescue.*

Hopi Indian Flood -*The* **3ʳᵈ Time** *was the time of destruction by water. The survivors sent out* **birds** *to test the land.*

Lakota Indians Flood-*There was worldwide destruction by flood.* **Birds** *were sent out by the survivors to test the land.*

Skagit Indians [Washington] Flood-*The creator made the world and 4 secret names. Only a few people should know the names, unfortunately, many learned them and became wicked. God made the world to flood. 5 people and* **some of each animal type survived.** **[This claim that the flood occurred because man had learned secrets he was not supposed to know is a common theme and should be considered as important.]**

Nisqually [Washington] Flood-The people became so numerous that they ate all the fish and game. They started to eat each other. They were **so wicked** *that the changer god, Dokibatl, flooded the earth. All living things were destroyed except one woman and one dog, which survived atop Mt. Rainer. From them the next race of people were born. They* **lived like animals** *until the Changer sent a Spirit to teach them civilization.* **[Possibly like monkeys]**

[Here we see gods teaching mankind, cannibalism, and wickedness causing the flood, just like the Biblical version, but I don't know about that dog, nor do I believe that a dog was the father of mankind.]

Let's see how the stories held to a similar theme. For this, I have provided a simple table of the most noticeable elements.

	Mid. East	European	N.Americaa	S.America	Pacific	Far East	Africa
Ancient humans were considered gods	X	X	X	X	X	X	X
Worldwide flood	X	X	X	X	X	X	X
Happened during the third age of man	X	X	X	X			X
It rained for 40 days.	X		X				
A human saved the animals	X	X	X	X	X	X	X
Few humans survived	8	9	8	4	2	1	2
birds tested the land	X	X	X	X		X	X
a rainbow was sent	X	X	X				
[other humans] survived	X	X	24	X			X
Humans became like monkeys	X			X		X	X

From the mountains of data we have, we can be certain that a worldwide flood occurred 9 thousand years ago, but evidence suggests that rain did not cause the flood.

Earth Shift Flood Theory

If the water didn't come from overhead, where did it come from? Here is one possible scenario. Remember from the previous section that around the time of the worldwide flood, the earth had just warmed up as part of its cyclic temperature change. Here is what probably happened following the stuff from the previous chapter as the ending of the Pleistocene would destroy almost all life on the planet.

*This increase in temperature caused the average depth of the oceans to rise as stated in the last chapter. Physical evidence suggests that a huge comet then struck the earth. Its high water content caused the ocean height to increase even higher but it still did not cover the entire earth with water as some have said. Finally; the Earth's axis shifted just as it had done a mere 3 thousand years earlier-either because of the comet strike or at about the same time. This time the shift was fast and furious. This action set the earth's water in motion. The fast shift in the axis caused some of the most severe tidal-wave action ever experienced. It was this action that covered everything and that is why the water is no longer on the earth. It never was here. [**The Bible called it a fountain from the deep.**]*

All these things occurred to initiate the worldwide flood. I know that's a lot of stuff to happen all at once, but evidence supports the events. Biblical texts and many other texts also indicate that something like a massive earth axis shift happened only 7 days before the flood due to a comet or massive meteor strike. We can be pretty sure that if and when the shift occurred, the previous poles would have quickly melted and refrozen at the new locations. If we look at the things that would happen we find 4 of them.

Ice turned to water, the comet added water, rain added water, water, and tidal waves occurred around the world.

The last one is the most difficult to accept in the religious world and its reasonableness and religiously acceptability will be discussed in a minute. First let's look at the first three.

Ice Turning to Water

Speaking of repositioning the poles, the huge ice blocks at the North and South Poles would have quickly melted when the Earth shifted. Let's look at the poles.

It has been estimated that there is about 25 million cubic kilometers of ice deposited in those areas. If we look at the Oceans, there would be about 250 cubic kilometers of surface to cover. Taking those into account, if the poles were instantly melted, the resulting water rise would be as much as 100 meters over the entire earth compared to what it is today.

After the shift was completed, the water level would slowly reduce back to its more normal levels as the "New Poles would begin refreezing the 25 million cubic kilometers of ice again. **That type of increase would not have covered all of the lands as suggested in most ancient accounts.**

Comet Water Didn't Flood the Earth

In addition to the 25,000,000 cubic kilometers of water generated by the pole shift, there would also have been the comet ice melt which certainly added to the already terrible situation, **but that also didn't flood the entire world or the water would still be here.**

Rain Didn't Flood the Earth

Just before the flood, as the earth axis apparently shifted by another 30 degrees, climatic instability would certainly have ensued as the Earth heating pattern below the clouds would have suddenly changed which initiated widespread raining. **This raining, of course, did not cause the flood**, nor does the Bible indicate that it did. The combination of rain and an ice to water transfer had certainly done their part to flood regions of the world and that will bring us to one of the most powerful and devastating forces on the earth.

Much of the Land Was Gone

From the Ice to Water transfer, The Increase in Rain, and the increase in water from a major comet strike, one might believe that as much as 500 meters of water increase could be accounted for. If you're wondering what the earth would have looked like just from a water increase of 500 meters, the picture below shows that only the mountain areas would have remained.

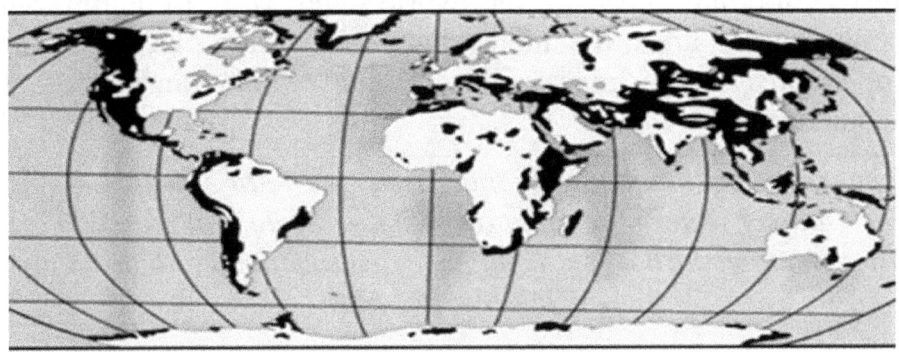

Only the Black areas in the previous map would have possibly been above the "normal water level" during the worldwide flooding. While the Earth land area was temporarily lessened greatly, there would have been a significant amount of area where people still could live. The Bible indicates that the entire earth was flooded, so either it is wrong or something else was going on. I say something else caused the eventual and total flood.

Tidal Wave Flooding

The thing that caused the rest of the flood was, most likely, tidal wave action unlike any we could possibly imagine. With the oceans already flooding most of the world, and with torrential rains flooding more area; the huge change in water placements and temperatures set up huge walls of water, which finished off the inhabitants.

*The water level had quickly increased by **100 to 200 meters**, but walls of water moving at an estimated rate of up to **1000 Kilometers per hour** would support waves estimated to be as much as a **kilometer high.***

These "destroyers" would have plunged their way around the new world. The very few people that were not drowned initially were trying desperately to find higher ground after the earth shift, but the tidal waves got the people that were left. These walls of water hit at different times around the world and destroyed everyone and everything. It, essentially, flooded the entire world.

Fountains of the Deep Evidence

The Bible might also give us evidence. Let's say you wanted to describe a tidal wave—a huge wall of water coming out of the sea--- you might call it a fountain of the deep just like the Biblical writers did thousands of years ago. Some indicate that there was a vast reservoir of water trapped in the earth that momentarily came to the surface and then went back into hiding. That doesn't make too much sense as any water trapped in that way would have superheated and caused significant surface notification. I think a "fountain from the deep" which could be translated as "pile of water rising from the ocean" could only have been a tidal wave. This is the type of thing I think is important to establish as one links science and religion together.

A Recent Large Tidal Wave

I know that a wall of water that high seems crazy, so I checked other tidal waves. Unfortunately most occurred when no measuring equipment was close by. The largest tidal wave recorded was on February 6th 1933 during a measly 100 kilometer per hour Hurricane. The USS Rampo recorded the monster.

The water speed was possibly only about 20 kilometers per hour. It produced a wall of water that was determined to over 120 feet high.

This was nothing compared to what happened in the past, but a wall of water taller than a 12 story building must have been impressive just the same.

Fossil FLOOD Evidence

Skeletal Flood Evidence

Near Lyons, France, the skeletons of 200,000 prehistoric horses are scattered in a small area. In a cave, in Monrovia, France there are enough mammoth teeth to fill a small sized hall. One thing that could have pushed these animals together during their dying time is a catastrophe such as gigantic tidal waves associated with a worldwide flood. Certainly there are other possibilities, but that one is the easiest to accept in my opinion

Rapid Fossilization Flood Evidence

Decades ago, Hugh Miller wrote that the entire British Isles are under-laid by billions of fish fossils, not laid down by normal sedimentary deposits, but many of these have arched backs distended gills, open mouths as if trying to get oxygen. Instead they were caught with silt in their gills. They were destroyed by the billions during some type of cataclysmic sedimentary event. They were encased is if a huge wall of mud slammed into them. This would most easily be explained by a huge flood and massive tidal wave action. Furthermore, the UK is not the only place to find these fish fossils; instead, thousands and millions of fish fossils have been found around the world. Many have retained all the body parts indicating a very rapid burial. Under normal conditions, fish do not fossilize. Dead fish are almost always torn apart by scavengers and disintegrated by bacteria. The fish fossils we see today were covered over and compacted very quickly as if by flood silting and mud deposition in massive amounts.

Whale Tail Flood Evidence

There are also huge ancient *whale fossils* that can be found. These too were completely and quickly buried in sediment. Near Lompo, California, for instance, there was found, in diatomaceous earth, an 80 foot long, Valine Whale upright on its tale. If you tried to sweep a creature like that up on its tail and then sedimentarily encase it, the job would require a global catastrophe and a large amount of mud. Make that a very, very large amount of mud. Unless someone can give me a better reason I'm going with a worldwide devastating flood here as well.

Planktonian Flood Evidence

"Geology" magazine in a recent article stated that Permian shales and chert in British Columbia show shifts in the Carbon 13 and Carbon 14 isotope ratios of decomposed plankton residue. According to the article, this shows that huge amounts of plankton all died suddenly within a few days. Two ways for that to happen is for it to be quickly yanked from it bedding during massive tidal wave action or for large quantities to be quickly buried in mud.

Dating the Flood

Ice Core Evidence

It seems like everyone and everywhere we are drilling into the icy glaciers to check on the conditions of the earth thousands of years ago. It also is apparent that the <u>percentage of Deuterium</u> in parts per million compared with normal hydrogen provides us with relative temperature changes in a region while the depth of the ice has been providing consistent and accurate dating. I'm not going to show all the graphs made from the dozens of studies, but it is interesting to look at two from opposite sides of the world. The first one is from Greenland while the second is from Antarctica. Low and behold, something very strange happened 9 or 10 thousand years ago and something again happened about 11 thousand years ago. While Greenland was getting colder during both of those times, Antarctica was getting warmer. I'm not going to say anything about my premise that the worldwide flood was 9 thousand years ago and a previous flooding event occurred 12 thousand years ago, you will simply have to make the connection on your own. I'm also not going to point out that the very sharp thermal spike during the world wide flood time may indicate that the earth axis may have shifted for a few months and returned to its preshift spin which would have caused earth flooding tidal-waves. The third thing I'm not going to point out is that the change that occurred 11 thousand years ago shifted the thermal average and has never returned to the old ones just like you would expect from a more permanent shift in the earth axis.

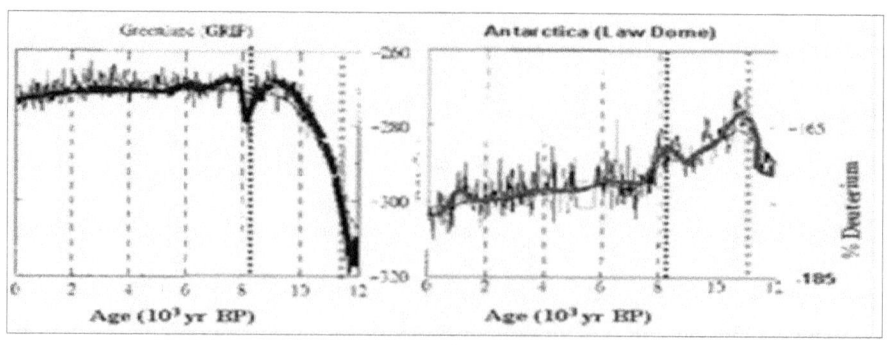

More Timing Evidence

In addition to written testimony there are substantial pieces of evidence, which confirm the destruction. The evidence below comes in the form of Oxygen, Sand, Nitrogen, Ice, Salt, the Atlantic, Antarctica, Mammoth tusks, crustal jerks, and a big old Crater. Some of this evidence also gives us an idea of the time-period. While these are simply short overviews of the results, the details of the studies can be found at various internet sites.

Oxygen Dating-The Dendrochronological Record shows a high spike in radiocarbons about **10 thousand years ago** which indicates that the ozone level was significantly reduced during a very short time period as would be expected from a massive meteor strike.

Nitrogen Dating- Increases of Nitric acid suddenly appeared in dated Ice core samples around **10 thousand years ago**. This shows that a significant amount of nitrogen in the atmosphere was burnt during that time period.

Ice Dating-The last Ice age retreated much quicker than previous ones to suggest that some major heating occurred about **10 thousand years ago**.

Sand Dating-Just above a layer of white 'sea" sand in Scotland, charcoal was carbon dated to be from over **7 thousand years old.**

Salt Dating- The landmass of Europe and Asia show significant evidence of an inland salt sea that has deposits dated to about **10 thousand years ago.**

Crater Dating- As evidence of a possible massive meteoritic storm, there is a giant crater near Bombay, India. The nearly circular 2 kilometer diameter Lonar crater, located 400 kilometers northeast of Bombay has been determined to be **10 thousand years old**. This crater could also be related to nuclear warfare of antiquity or a massive meteor strike. No trace of any meteoritic material has been found at the site or in the vicinity of the site, but there is a possibility that most material could have been melted by the event. There are indications of great shock and intense, abrupt heat, which caused glass spherules to be formed. These bits of glass are the only remains. The hole is not large enough to have been the main part of the comet we are searching for but the timing is about right to have been one of its pieces. By the way, a comet would not leave as much meteorite evidence as other falling bodies, because of the high concentration of ICE.

Atlantic Basin Dating-In the middle of the Atlantic Ocean, the crust is continuously opening as the Atlantic gets wider every year. As part of this expansion, volcanic material is deposited. By looking at the alignment of the metals in the material with the "current" magnetic pole position, it was determined that there was a major shift in the magnetic pole **about 10 thousand years ago**.

Mammoth Tusk Dating-Preserved Mammoth tusks and teeth have been found in ancient layers of peat off the coast of America have been carbon dated to be about **10 thousand years old.**

Antarctica Dating-Antarctica was possibly one of the commercial centers of the world before this event. It probably moved to the South Pole and, of course was destroyed. As proof, fossilized plants and animals have been found in core samples, which indicate that the date when Ice "last" appeared on the continent was about **10 thousand years ago**. [Several cycles have been found of colder and warmer climates on the southern Island.]

Crustal Jerk Dating-Physical evidence found around the world shows signs of crust displacement. It seems that the "crustal jerks" occur in spurts. The current hypothesis is that "Crust displacements" occur every 40,000 years and the last one occurred **9,000 to 12,000 years ago**. [What the many researchers in this area mean by crust displacements can be generally explained by the movement of the axis of the earth. Whether the outside of the earth shifted or the entire earth shifted is not relevant to those living on the outside of the crust.]

Pleistocene

Finally the Pleistocene Extinction occurred 10 thousand years ago and the floodwaters, the earth shift, the massive tidal-waves and cosmic rays barreled down as the atmospheres changed and the earth shifted to a new rotational plane.

Earth Sift Evidence

Uplifted Mountain Evidence

Let me quickly discuss another indicator that has perplexed many people. Even staunch observers of the plate tectonic mountainization theories don't like this area. If you remember from earlier, I said that the plate tectonics brought out in our science books to explain the mountains of the world was not a correct theory and that Mars, most likely, caused the "uplift" of most of the mountains of the world. While that does explain many of the mountains, there is at least one area that was not mountainized that way. The reason this particular area should give us so much concern is that the indications are of a great mountain making movement that occurred very recently. For the story we go to Peru.

Peruvian Mountain Lifting

Tiahuanaco is a Pre-Incan city that was built about 15000BC as determined by something called its "Kalassaya stone" which creates an alignment with the equinox sun at that time. [A second researcher estimated this time to be 10,000 BC by another star alignment so you'll have to take your pick]. No matter when the city was constructed, it suffered immense damage about **10 thousand years ago.** Earthquake-proof walls were destroyed and the people simply disappeared. Even though the damage could have been from the many wars, it is more probable that a great catastrophe like the earth shift actually caused the destruction because there is another significant indicator. Tiahuanaco was built as a seaport.

This seaport is now 12 thousand feet above sea level.

In addition to the port city, complete with docks and the other necessities of water trade, there are three huge "salt water" lakes in the same area and isolated 12 thousand feet above the rest of the Pacific Ocean. The lakes are Titicaca [4 thousand square miles in area and as much as 900 feet deep], Coipasa and Poopa [both 200 square miles in area]. All the lakes reside in the same area. I think we can assume that the immense damage of the city and the 12 thousand mile pushup that land locked portions of the Pacific Ocean occurred at the same time, which places the incident right in the middle of this earth shift thing. The shift must have been extremely violent. Alaska and Siberia also show the effects of this violence, but in a different way.

Alaskan & Siberian Evidence

Many parts of the Earth were quickly frozen by the move. Winds were like tornadoes. Mountains of mud covered the dying animals in locations with the most extreme changes such as in Siberia and Alaska. Below are some of the current findings in those "hard hit" areas. Imagine what the people at that time must have thought. According to ancient writings, this was God's way of waking up humans and angels about his power, but the event was quickly forgotten by the descendants of the survivors.

Alaskan Evidence

Hundreds of thousands of Mammoths and bison were found torn and twisted. Portions of the flesh, toenails, & hair were still on blackened bones. The neck and skull of a bison vertebra was found still clinging with ligaments. Mixed with the huge piles of bones were twisted trees. As a point of reference, there were 1766 jawbones of extinct bison found. These animals became extinct about 9 thousand years ago.

Siberian Evidence

Siberia is a frozen, twisted, mammoth grave-yard. Food found only in temperate climates was still in the mouth of one of the victims and over 20 thousand tusks every decade have been brought out as bodies are found. Just like the event identified in Alaska, this was no ordinary catastrophe. These animals were violently smashed together and quickly frozen as the earth's axis shifted.

Other Areas Moved North

Other areas weren't left out of the destruction. Here are examples from California, Nebraska, Germany and even Florida.

Nebraskan Evidence

Smashed piles consisting of thousands of bones of extinct rhinoceros, clawed horses, and giant pigs were found as though they had been violently killed together at the same time.

German Evidence

Hippopotamus, ostriches, seals, and reindeer were all killed together in an area of Germany and their bodies are all in one area.

Floridian Evidence

A pile up of animals has been found. The death bed was found just south of Tampa. It has, so far, yielded the bones of more than 70 different species of animals, birds, and aquatic creatures all jumbled together. About 80% of the bones belong to plains animals, such as camels, horses, mammoths, bears, wolves, large cats, and a huge bird. Mixed in with all the land animals are sharks' teeth, turtle shells, and fish. The bones are all smashed and jumbled together. The mixing of the aquatic and land animals in this way suggests a huge group of tidal waves not unlike that expected from a crustal shift.

Californian Evidence

In the LaBrea pit, hundreds of saber-toothed tigers, horses, bears, lions, camels, mammoths, mastodons, bison, and peacocks are all pilled together like a violent situation occurred. These animals were twisted together because of some pools of tar.

Antarctica Moved South

Recent core samples from Ross Sea reveal a time when Antarctica was not cold. Ok, it was cool, but not the chunk of Ice it is today. Abruptly the core indicates glaciation occurred and continued for thousands of years. The time for the last transition has been estimated somewhere between 8000 and 5000BC. My personal belief is that it occurred about 7000BC when the rest of the ruckus was going on.

Flood Problem

Ok! I've described to you elements that show there was a massive flooding 9 thousand years ago, but there is a problem that I do not have an answer for. People have found hundreds of Mammoth, saber-toothed cats, Rhinoceros, and other animals that have been tangled up in huge piles of what are now massive bone-yards all over the world. The bone piles point to a horrible change in the earth and an instant destruction of almost all living things. The problem is significant in its blankness. THERE HAVE BEEN NOW HUMAN BONES FOUND IT THESE AREAS. If we assume that millions of people met their fate in the last major cataclysm of our world, we should have found some large piles of human bones as well. Perhaps we will find them some day or possibly they have already been found and the secret has been well guarded as a place of terror. Whatever the reason, I'm with you wondering how the evidence suggests that no people died in the flood that killed everything else. I'm not saying they didn't die, I simply saying that the data doesn't support the massive loss of humans. Life. While the physical evidence is limited, there is certainly something that is not. That "something" is written testimony of the catastrophe. Some of the written testimony even affirms the dating..

After the Flood

The evidence and written testimony tell us that mankind immediately following the flood episode was different than we are today. For one thing, there was quite a bit of knowledge passed on from before the flood. I don't mean a little bit of knowledge passed on from Noah; I mean huge amounts of knowledge survived the floodwaters. Much of it was eventually forgotten and that's where the Tower of Babel will come in. Right now let me just give you a quick overview of the knowledge of mankind before Babel. They were adept at genetics, aeronautics, medicine, and many other sciences we are relearning today.

Societies were built up way too fast for the "Normal" civilization process and people today don't seem to recognize the time problem, but it will not simply go away just because we don't think about it. We know that the world was destroyed by flood due to physical evidence around the world and that the flood occurred 8 to 10 thousand years ago. What we are finding is that civilizations sprang up as though they were already here for thousands of years.

Some people use this miraculous civilization expansion as proof that no flood ever occurred, but I believe from the evidence presented here and other evidence you may know, that there can be little doubt of the catastrophe. The elements of civilization must have been brought with the antediluvian refugees. Not only was there one civilization that sprang up overnight, we can find many instant societies around the world as though there were a number of survivors of the flood. Here are a few examples of the strange capabilities that these "ancient" societies had. Capabilities we are only now beginning to understand again.

Sumerian Instant Technology

Somehow, almost immediately after the flood, the Sumerian civilization went from nothing to a highly technical civilization within a very short time. This could only have been done if

179

survivors of an advanced race of people before the flood initiated and taught the inhabitants. By teaching the people in the ways of science and technology, the surviving "gods" insured that they could take control and were worshiped as god–kings. Some may want to discount the "gods" statement, but the ancient texts tell us that the rulers were worships as if they were deities. This is an impressive list of accomplishments from about 6 to 7 thousand years ago.

Brain surgery, cataract removal, and bone scraping techniques, fine medical instruments, herbal medicines, beer enemas, and alcohol as a disinfectant, all made their medicine fantastic.

Models of various organs were used in medical schools.

Artificial gemstones, textile manufacture and basket weaving were common.

Lenses- From Turkey to Mesopotamia, there have been found a quantity of about 75 Plano-convex glass lenses. Here's an even more strange part. In at least one of these lenses was found cavities that could have been just to introduce some type of gas or substance to change its optical characteristics. Who knows what they might have been doing with these "modern things".

Use of petroleum products for fire, heat, asphalt road building, and waterproofing were all known.

Metallurgical technique included kilns, working bronze, firing metal, casting, and finishing copper mirrors have been verified.

Robots, Laser Weapons, and flying machines were all mentioned in ancient texts.

Irrigation, grinding cereal into flour, fermentation, cheese, bread loaves were in use; especially the good old fermentation.

Pottery, ship building and chariot building, and the banking industry showed a high level of society.

The first bicameral congress, code of laws, and a 4 judge justice system were all quickly introduced.

The first historians and the invention of writing are attributed to this time.

Required schooling- Their school used a novel idea; it was called "man in charge of the whip" and there was not as much misbehavior in the classrooms.

Pregnancy Testing- Ancient tablets tell us that insertion of a woolen "tampon" could be removed and treated with "alum". If the wool turned red, the woman was pregnant.

Large libraries were established [30,000 texts found] and groups of proverb texts. Here are some that are still true today.

- ***"If we are doomed to die***, *let us spend. If we shall live long-let us save"*

- ***"He that has silver***, *may be happy. He that has much Barley, may be happy; but he that has nothing can sleep."*

- ***"Man: for his pleasure***: *marriage; on his thinking it over: divorce."*

- ***"In a city without watchdogs***, *the fox is the overseer."*

Ancient Africa

In Africa we find something strange. People living immediately following this worldwide flood have technologies that greatly exceeded the technology some 5 thousand years later. Let's see what technology was like in Egypt and central Africa.

Egyptian Instant Technology

The Egyptian did the same thing as the Sumerians. Again, this could only have been done if survivors of the advanced race of people from before the flood initiated and taught the inhabitants. According to the ancient "Emerald Tablets" God sent some of the "survivors of Atlantis" [angels] to teach the Egyptians and build the pyramids. Just like in Sumeria the "antediluvian gods" and their offspring [demigods] were treated as god-kings. By the way, These gods are the same thing as the Creator God, they just remembered things that made them powerful and useful.

Herodotus wrote about the fantastic Egyptian Doctors. He said that each doctor specialized in one type of disease. He also said, "Physicians were in every community. Some would only work on eyes, others the teeth, still others the intestine or internal organs." He also talked about the common use of brain surgery.

The Edwin Smith Papyrus, estimated to be over **5 thousand years old,** has 43 sections just on the treatment of wounds and fractures categorized by body area. Each section contains information on examination, diagnosis, treatment and prognosis of recovery.

The Berlin Papyrus discusses the **removal of cataracts** from the eye. The papyrus is believed to have been found under a statue built during the reign of King Sent who reigned 6 thousand years ago. There are no incantations in the work, as was the custom for later medical documents, so there is secondary cause to believe in its extreme age.

Brain surgery [drilling holes into skulls] was done successfully, just like it was done in Sumeria.

Prosthetic parts have been found; at least wooden toes were made and worn. Check out the one to the left. It has its own toenail secret compartment. The one below right doesn't have a secret compartment, but it shows that prosthetics was common and loss of the big toe was common as well.

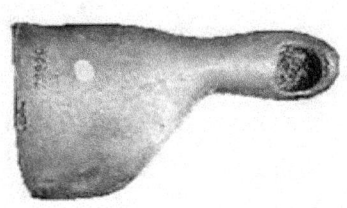 

As far as chemistry, they must have been very much into that particular science as the name comes from the original name of the Egyptians; Khemites. They had **contraceptive jelly** and urine pregnancy testing methods.

Their mummification was best in the world. OK! It was only best because the desert did most of the work, but they were still good at it.

Molding solid diorite stone [hardest known rock], was evident.

Levitation of blocks was apparent. The method was written about and heavy pieces of evidence are everywhere.

Metallurgical techniques [electroplating antimony and gold] were accomplished.

Models of airplanes have been found.

There is strong evidence that many types of electric devices were used.

Complex irrigation and making beer were both done.

Ship building [in the desert] has been evidenced.

Great historians and early forms of "complex writing" were established.

A **society of magic** was revered.

The picture below is of a gold tooth-plate, which shows something about the level of dentistry they had developed. Gold bands were also used to support good dental operations as shown on the right.

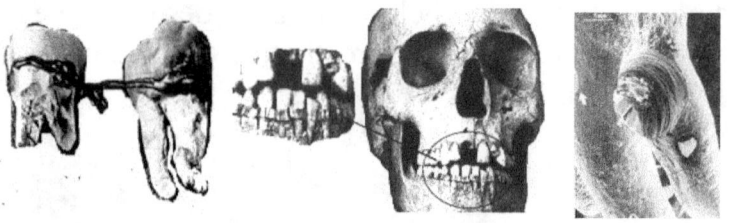

The far right shows a hole drilled in a tooth, the reason for the drilling is unknown.

Egyptians Invented Aspirin

In Egypt was found a remarkable artifact. Now called the Ebers Papyrus, it is a collection of 877 medicinal recipes from about 4 thousand years ago. One of its curative concoctions is made from dried myrtle leaves for rheumatic and back pain. Guess what is in myrtle and willow leaves. You guessed it Salicylic acid [the main ingredient of aspirin.] By 400 BC Hippocrates of Kos, the father of all doctors, recommended a tea extract from the bark of the willow tree for fever, pain, and labor, as he must have studied the Egyptian ways. The news spread and in China and Asia, and among North American Indians and the Hottentots of South Africa the beneficial effect of plants containing salicylic acid well before its rediscovery in 1837.

Egyptians Treated Tumors

That certainly wasn't the only 'normal' medicine or curative element known by the Egyptians and recorded in the Eber document. They also knew about tumors and how to treat them.

In the Ebers Papyrus it says, "a tumor affecting a vessel which is stone-like to touch is suitable for surgery---cauterize the wound lest it bleed too much."

Central African Instant Technology

African Medicine, Astronomy & Weapons and their knowledge of the universe was unbelievable. In addition to the unbelievable knowledge that the Dogan gained about astronomy, there were other anomalies. Here are some of the more unusual capabilities and elements of knowledge that have been reported from about 6 thousand years ago.

They knew that the star Sirius had a twin star Sirius B, which was only rediscovered recently.

They knew about two of the moons around Uranus without the aid of a telescope.

*Several skulls were found in Rhodesia. In the skulls are round holes with no sign of radial fracture at the entry point- shattered on the exit. Arrows couldn't cause the holes nor any other weapon besides the high speed projectile from a gun. The skulls are estimated to be well over 6 **thousand years old**.*

Ancient China

The Far East also gained knowledge too quickly. The Chinese ancient society had some of the most extensive knowledge in medicine. They also remembered other things.

A Medical Document from before 300 BC *stated the following—* *"He gave them a toxic drink & they were unconscious for three days. Then Chao opened their stomachs and examined their hearts. After removing and exchanging their organs, he gave them another drug and they went home recovered."*

For smallpox inoculations, *the Chinese knew to rub dried matter from small pox sores onto the nose of unaffected people to insure immunity.*

The "Book of Medicine", written about 2650BC, *contains this description, "The blood is controlled by the heart and regulated by it. It flows in a continuous circle and flows ceaselessly."*

Chinese *were the first to remember the art of making gunpowder, and paper, and flying, and the compass, and earthquake sensing devices, and many other things which they should not have known how to do given what we think of as normal scientific advancement.*

Ancient Greece

The ancient Myths may be more than just obscure ways to train social etiquette as has been determined today. They may tell us about some of the ancient capabilities in Greece.

Greek Robots- According to Greek Mythology, Talos was a giant bronze robot created by Hephaestus, who was programmed to protect an island from strangers. Because he was made of Bronze, he was practically invulnerable.

Rejuvenation of Life- For this story we go back to Jason of Argonaut fame and his "god" lover, Medea. They found out that King Pelias had killed Jason's parents, so Medea promised the king she would make him young. Now if that doesn't seem right we need to read the rest of the story. She took an old ram, cut him into pieces, boiled him in a cauldron, and then took a young lamb out of the cauldron. All Pelias had to do is let his daughters chop him into little pieces and Medea would boil his pieces to make him young. **Oops!** The boiled pieces stayed boiled. [**The reason I brought this up was not to demonstrate stupidity, but to show that ancient people believed in bringing people back to life.**]

Other Marvels- The early Greeks also provided us with descriptions of invisibility helmets and rings, flying shoes, lightning weapons, and many other modern conveniences. By the way, this whole "Argonaut" episode [which means "Heavenly Ship"] sounds more and more like a star ship the more you read the Jason story. Why would there be so much emphasis on a "Golden Fleece" protected by a dragon anyway? It could be that the myth is less fantasy and more reality.

Ancient Scotland

In Scotland was found the remains of an ancient town of 7 buildings called "Skara Brae". This town is very unusual in that the construction was apparently was at a site that had minimal resources for food, so food had to be brought in. The apartments were buried halfway in the earth, regularly shaped, and very efficiently laid out with stone drainage systems. As shown below an apartment had one or two beds [which are believed to have been covered with a canopy], an easy chair and hassock made of stone, a central open pit hearth, and an inside water tank packed with clay; all the comforts of home. The town mysteriously arrived at over 5 thousand years ago and was completely abandoned around 2600 BC. **[This was during the Babel Wars and the people in this remote spot were there because they were afraid.]**

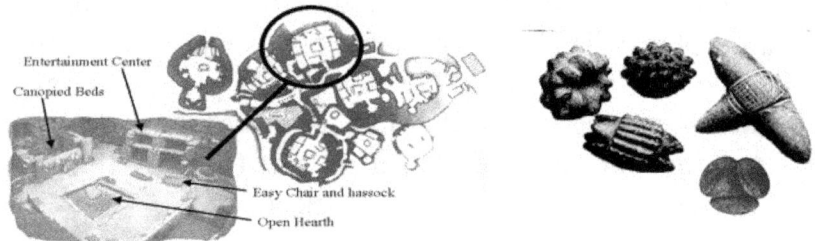

Money-In Skara Brea, the idea of money must have started quickly. Hundreds of these unusual "balls" have been found which are estimated to be over **5 thousand years old**. The "balls" are in many shapes and have been worn down as if they were handled regularly, but are not chipped as if they were used as a sling projectile.

It is believed that these unusual items may have been the first non-barter money items. Here is a picture. What do you think?

Ancient India

5000 BC City Construction

Because of its location near the ancient Garden of Eden [which was found to have been located somewhere in Afghanistan] and the early settlements of the Adamic humans in the area, Pakistan must have been at the center of many trade routes between 5 and 7 thousand years ago and, just like the other places, civilization quickly sprang up in the region. Mohenjo-Daro has been determined to be the first city that was built in a very modern way. The streets were all laid out in a grid pattern and there was underground sewage. Throughout the city, widely separated sewage and water lines, were laid out in the same pattern showing the great understanding of sanitation requirements and that the city itself was fully planned before it was built.

A drawing of what it might have looked like is shown below. Also, photographs of the remains of the city and the covered sewage system are shown. The capability to plan and erect a city like this must have come from remembered knowledge from before the flood. This knowledge, like many of the other things, was soon forgotten.

Naked Civilization

Not only did the city have its central bathhouses, public arena, and other signs of highly developed civilization, it also had its dancing girls and artists to show them off. Here is a sample. Like most of the women statuettes that have been found, this one was naked, of course.

 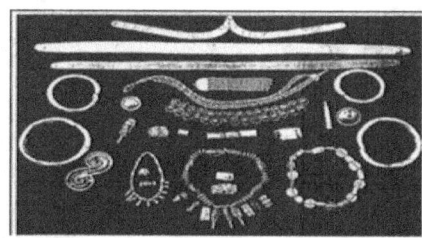

Jewelry-ated Civilization-All types of intricate jewelry found in the area shows a high level of civilization, as well.

Instant Technology in India

While I'm not going to get into this fact too much, the ancient Indians knew about flying ships better than just about any nation. Not only was it the center for knowledge about flying machines, but also many other medical wonders were accomplished and recorded there.

*The "Samhita", written **8th century BC** contains specification for transplanting tissue.*

Ancient Indians would rub the crust from smallpox sores on to open cuts to immunize people. Just think what the dark ages would have been like if the Black Death, smallpox, had been cured by this technique or the Chinese technique in Europe a thousand years later.

The "Sushutra" described many surgical techniques including removal of neck tumors, Tonsillectomy, and amputation techniques. It describes proper use of over 121 surgical tools and had a detailed description listing of many diseases of the day.

"Celus" confirmed ancient transplanting techniques used by the ancient Indians during Roman times. Transplanting skin to manufacture a new nose whenever one was cut off in battle included partially cutting off portions of the skin from an arm and attaching the skin to the face while still attached to the arm. This allowed the graft to retain blood-flow during healing.

Ancient South America

The advancements were not isolated in the Middle East. Pictographs on as many as 15,000 stones have been found indicating all types of advanced technical knowledge of the pre-Inca civilization from over 6 thousand years ago. On the plains of Nazca, Peru hundreds of line drawings and long straight lines similar to airport runways and a golden model of an airplane was found. A temple city was made on top of the mountains with perfectly cut 500 ton stones that fit so close together that a knife cannot be forced between them. These and other unbelievable marvels could not have been the work of anything less than an advanced civilization. The remains of the antediluvian society during what the Inca refer to as the age of heroes were responsible. Here are some of the curious capabilities.

Herbal medicines and blood transfusions were used.

Possible organ transplant technology was practiced.

The ability to work with Platinum was known.

Cesarean section birth, acupuncture, and anesthesia were all here.

Cement was used for filling teeth that still is holding in the skulls found today.

Pictures of men dominating dinosaurs were shown on some of the 15,000 writing stones that were found

Huge pictures were drawn in the Nazca plains that can only be seen from the air.

Knowledge of telescopes was shown on stone pictographs.

*Knowledge of the world as it looked over **5,000,000 years ago** can be seen on pictographs.*

Vast knowledge of astronomy with an extremely accurate calendar that shows an end of time or destruction on December 12, 2012 shows the high level of mathematics.

Flying machines were apparent.

<u>Brain surgery</u> was commonly practiced and people survived. Hundreds of skulls have been found with surgically administered holes. Over 85% of the skulls show that the patient survived a long time after the procedure. One skull was found with 4 holes drilled. The patient lived after all surgeries and the skull opening had healed. Below are a couple of examples one with three easy round holes and the second with a chiseled out square hole. Both patients lived a long time after the surgery.

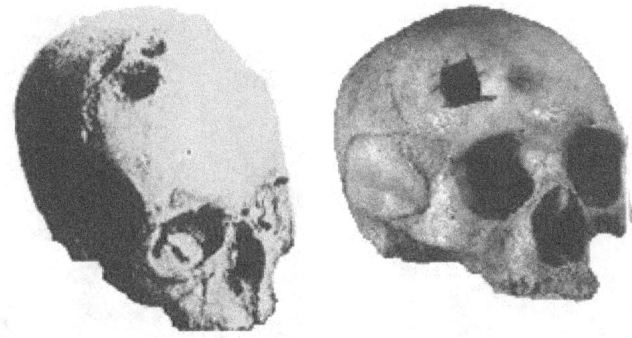

Instant PreInca Machinery

All over the Peruvian landscape can be found what appear to be the remaining mountings of some intricate machinery. Adding blocks to see what it may have looked like before the other parts disappeared doesn't really help. The drawing below left surely shows the remains of an odd contraption. Of course, we don't know what the machine really looked like because all that is left is the outside stone structure, but the blocks I have shown show the multitude of regular geometric shapes that defined the structure before the metal-works had finally disappeared over time.

The second image, is of some of the many gear found. We don't know much about their machinery, but what they left has everyone scratching their heads. The image below again show the mounting points for another machine, but the last image is really neat as it appears to be a complete 3D map of some city. As far as we know they didn't have mapping software, so this could have been a challenge.

The Tower of Babel

Some of you may be shocked to find out that there was a real Tower of Babel. The building of the Tower of Babel Citadel and the events that surrounded this Biblical icon were the last major events to mold the earth and civilization. Not only did this building or buildings collapse, but also the world and the people living in it were changed forevermore. In a nut shell, here is what the evidence tells us.

There was a huge war about 6 or 7 thousand years ago. It was worse than any in recent memory.

As part of the war a major launching site of some kind was designed and built. The citadel making up the launch facility and other components of this area were known as the Tower of Babel.

As an aside, this "tower" was not located in Babylon as you have been told. The only reasonable candidate for the Tower building is located in Lebanon; in a place called Baalbek. While I don't go into the enormous amount of evidence to support this theory in this book and its location is not important to earth's development, the misrepresentation of its placement shows that we are told "Historical Facts" that are more convenient than factual.

Nuclear weapons were ignited around the world. Surely you were told about the massive war times that resulted in the more devastation than all modern recorded wars combined.

Thousands of scared people began living underground around the world. Some cities go underground by 20 or more stories and still this terrible time was not recognized as important to many historians and earth scientists.

The wars were horrible. Someone was using germ warfare or something similar during the war and the outcome was devastating. No! It was worse than terrible.

Outcome of the Babel Wars

You have all heard about the Tower of Babel and the fact that people could no longer talk to one another. Besides the Jewish version, similar accounts can be found around the world. The destruction of Babel was termed as a punishment by many religious and historical documents. Most talk about how men became like monkeys or apes. [Hopefully you picked up on some of that in the section on the worldwide flood.] I know that this whole concept may seem absurd, so the first questions might be, "What really happened to humans after the Tower incident and could humans have had a single language at one time?"

The Single Language of the Masses was lost.- Let's theorize that man may have been able to talk without words, telepathically in the olden days. If telepathy was the communication used, human beings could think about what they wanted to convey and presumably, everyone could understand this type of "thought" language. If this whole telepathy thing has validity, then after Babel it became difficult for man to understand the patterns emitted from the brain or brains could no longer transmit thoughts. I know some of you are giggling at the telepathy thing, but let's not discount telepathy for a moment. Let's first see what the Bible has to say about Telepathy. This time let's carefully read the actual words.

Genesis 11:1- *"Now the whole earth had one language and few words."* [Notice it says that people used very little spoken words. The only two ways that could be possible. One is that people had little to say, but the more probably idea is that most communication was done without words.]

Almost everyone that survived in the entire human race was affected physically and biologically including those people living underground. Huge amounts of evidence tell us the same thing, Their DNA structure was affected by the manufactured germs or some other catastrophe which carried the "infection" to all of their children.

Humans could no longer use most of their brains and many capabilities they once had would be lost forevermore.

Human life spans decreased drastically and instantaneously around the world.

As many as 1/3 of the entire population of the world died as a result of this horrible war period according to ancient texts and physical evidence.

Humans lost telepathic powers as a result of this incident. On both sides of the Atlantic, the same story is told.

Here is where something modern mathematicians call super-symmetry comes into play. It states that if someone modifies mass in our universe, there is an equal and opposite effect in a parallel universe. Nuclear change of a particle could cause this affect. --- anyway!! While the war was not generally waged between people in the "parallel universe we call heaven" and "our universe", we are told in the Bible and other ancient texts that the heaven groups were affected enough to do something about the war. The Biblical story tells us that the Tower was toppled and humans became like monkeys. [There are great details in the book of "Jasher".] The germ identified above may have already sealed Human fate even before the tower was blasted. [While I'm sure you are doubting this drawn out statement, the evidence is enormous.]

If you wondered why God would have been so worried about a measly building that he destroyed the Tower of Babel, this theory will help.

If you have noticed like many archeologists that people seemed to have been more civilized 8 thousand years ago than they were 4 thousand years ago, this theory will help.

If you have ever wondered why human brain size expanded very quickly during his "development" only to shrink during the last few thousand years, this theory can help.

If you wonder why we only use a small part of our brain, this theory will help.

If you have ever wondered why the Bible talks about people losing the capability of communication, but the people could still talk to their family members, this theory will help.

Everything can be explained in a scientific way that does not disregard our sacred and ancient religious texts.

Brain Atrophy

Many have presented the premise that we only use 10 percent of our brain. Here is another simple observation that has to do with unused organs becoming atrophied or completely disappearing. Whatever made humans not use all of their brains did not happen hundreds of thousands of years ago, but instead must have occurred only thousands of years ago. Because the loss of brain use happened only recently the brain size has not had time to shrink much. This shrinkage would be kind of like how modern human's appendix is substantially smaller than older versions of human's appendix because we no longer use it. While it hasn't shrunken a lot, it appears that the human brain began to atrophy immediately after the Tower of Babel incident.

Brain Size Evidence

For a minute let's look at the brain sizes again along with the average dates for various humanoids. For this study we start with the Australopithecine Africansis named Lucy and go to modern Cro-Magnon humans. The line at each site represents the largest and smallest brain found from each of the subgroups. Its horizontal position dates the meantime existence of a particular species. Generally speaking, the brain size got larger and larger then, all of a sudden, it shrank sometime between 50 thousand and 4 thousand years ago. **It doesn't make sense unless something happened to our genetic messages that formed our brain.** I wouldn't be surprised to find out we even lost the ability to speak telepathically because of this reduction in capacity. Please also note that the data is even more significant in that Modern Man is taller than Neanderthalis man so the brain to size ratios even are more skewed. Our brains are getting smaller.

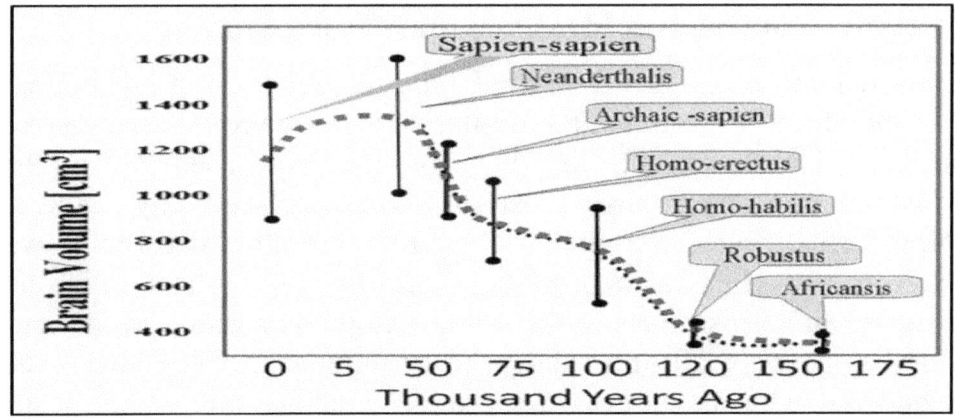

Humans Dumbed Themselves

Ancient works tell us that God caused the loss of communication. The texts indicate that he punished the builders by talking away their speaking capability, but that answer doesn't fit the way God does everything else so there must be more to it. The more reasonable answer is that natural order was responsible for the catastrophe and God defined the natural order of things so in a roundabout way he "dumbed the humans". That doesn't mean our creator couldn't zap us into or out of existence, it just means he always seems to work within the physical environment he has set up.

Humans Designed a Bug

That being said, why does all the evidence point to the inevitable truth that everyone seemed to lose intelligence after the "Tower Incident" and why are our brains almost totally unused today? One possible answer is that during the process of humans investigating and genetically altering animals, they probably made an animal that they didn't want. Today we have this same problem and have released harmful viruses and other animals that were created in a laboratory. The same thing, apparently, happened over 5 thousand years ago.

Bug Theory

Possibly, this germ that was genetically engineered during the Babel War essentially ate the enzymes necessary for the brain to function completely and the germ still does the same thing today.

No One Has Identified the Missing Enzyme

I know that no one has found this type of limitation in our brains, but that doesn't mean that the limitation is not there. There must be some logical reason why our huge brain is only used in a minor way today. There also must be some reason why ancient Neanderthal brains were even larger than the atrophied brains we have today. There also must be a reason why ancient religious leaders and historians told about how communication and intelligence was, all of a sudden, lost to humans. If you have ever wondered why an idiot savant can do amazing things that seem impossible, the answer is that his brain was affected differently by the "germ" in one area and we can get a glimpse into our past capabilities.

Brain Atrophy

The answer may be that a brain germ made MOST of the brain useless and the useless parts simply began to shrink away.

I know you are having a hard time with this whole concept, but the facts seem to indicate this very important point when trying to understand the crazy development of societies on the earth..

Don't Disregard This Probability

Not believing this is a possibility doesn't make it go away. As I previously described, civilizations emerged from the worldwide flood with ancient knowledge from before the flood and set up societies that almost immediately sprang into a high level of civilization including exotic medical treatments, highly advanced manufacturing techniques, world trade, holographic and optics knowledge, knowledge of flying and many, many other things BEFORE the Tower of Babel destruction. There are so many examples of these advancements around the world, that **this** truth is highly probable. Just as obvious to researchers is the fact that, all of a sudden, people around the world became ignorant. I'm not talking about religious dogma, I'm talking about physical evidence. No longer did people remember the exotic medical techniques, building methods, use of optics, knowledge of flying, or just about anything else. The physical evidence is ample but there are historians that try to smooth this awkward transition by ignoring the predecessor capabilities of mankind and smearing the limitations after the transition. It simply doesn't work. Another problem for historians is the length of life enjoyed by the PreBabelic people.

Life Was Shortened

It is my belief that the actual shortening of life did not happen because of some water, increase in cosmic rays, inbreeding, or genetic decay. I believe it happened when the loss of communication was noted; when the loss of long distant viewing occurred; and when the loss of "knowledge of all things" happened and when many humans began to take on the apelike traits. That time, by most ancient sources, was the time of the Tower of Babel, some 6 thousand years ago. If there was a major war, the probability that germ warfare was used is highly likely. We use them today and it is a very effective way of taking control of areas without much bloodshed, but here's the downside.

The likelihood that something went wrong with some of the bugs also seems probable. If these bugs could mess up DNA and not let us use some of our brain as is stipulated and/or easily inferred from ancient documents and other evidence, this same "bug" could also disrupt the DNA sequencing such that life-times were drastically shortened.

Whether you believe that the preflood humans had lifetimes of a mere 1 thousand years or over 5 thousand years as suggested below, it doesn't matter when arguing this point, because there was a major change 6 thousand years ago and the Tower came down about that same time. The first table is from Jewish and Biblical texts. The antediluvian ages have been expanded by a factor of 5 to keep in line with other evidence that suggests the Adam creation occurred about 40 thousand years ago. It also was modified to place the worldwide flood at about 9 thousand years ago as the physical evidence suggests and was presented. Please notice what happens after the Tower incident. On the left are the patriarchs of King Lineage and the Blue blocks show the lifetimes of each.

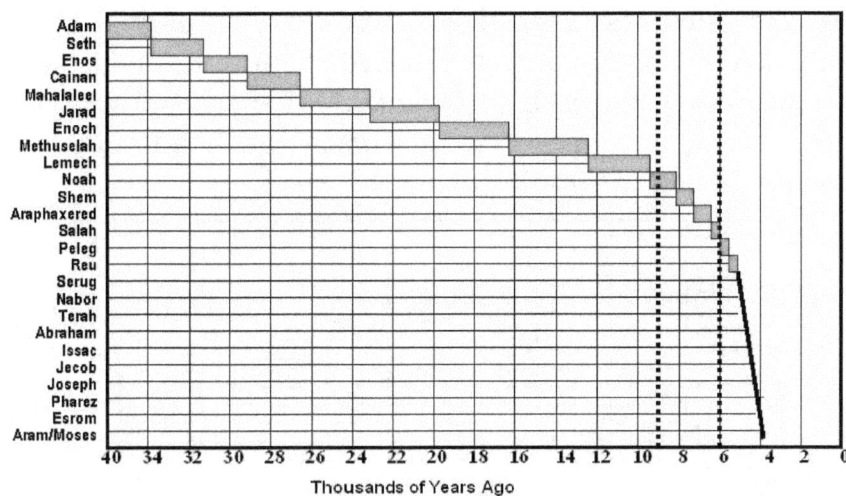

Thousands of Years Ago

The Chaldeans and Babylonians had a similar structure of their historical timeline. In this one the King's reign time periods have been reduced to 1/12th the recorded time in keeping with the 40 thousand year common point. Again we find a drastic reduction in ages presumably after the Tower of Babel incident. In the graph the blue blocks are now representing the King's reign time.

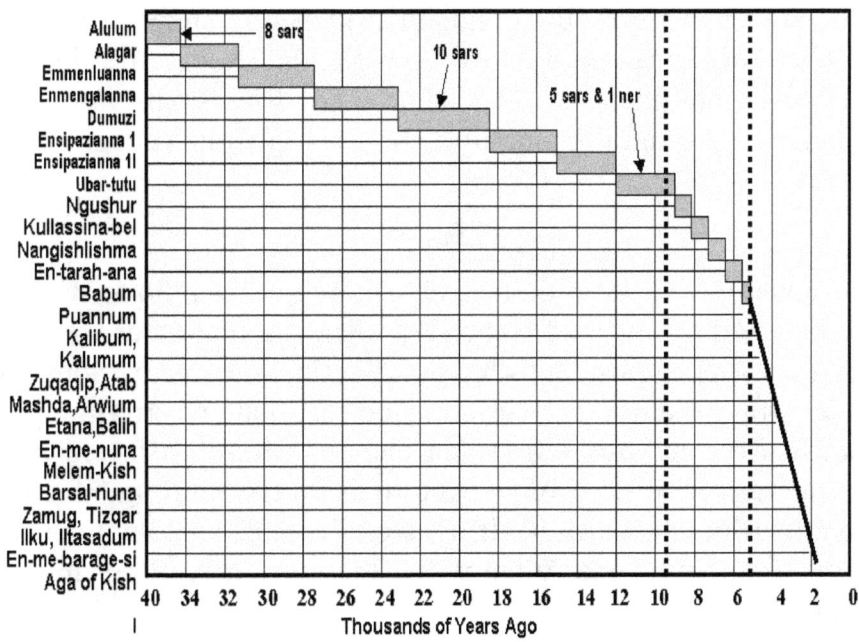

Thousands of Years Ago

Like the others, the Egyptian timeline looks the same with long reigns before the flood and tower, followed by much, much shorter reigns and lifetimes after the Tower "war". In this case, the antediluvian reigns have been reduced in time by a factor of 6. Rather than showing each of the rulers after the Tower, I have clumped them by Dynasty so that each block after the Tower represents about 6 to 10 rulers.

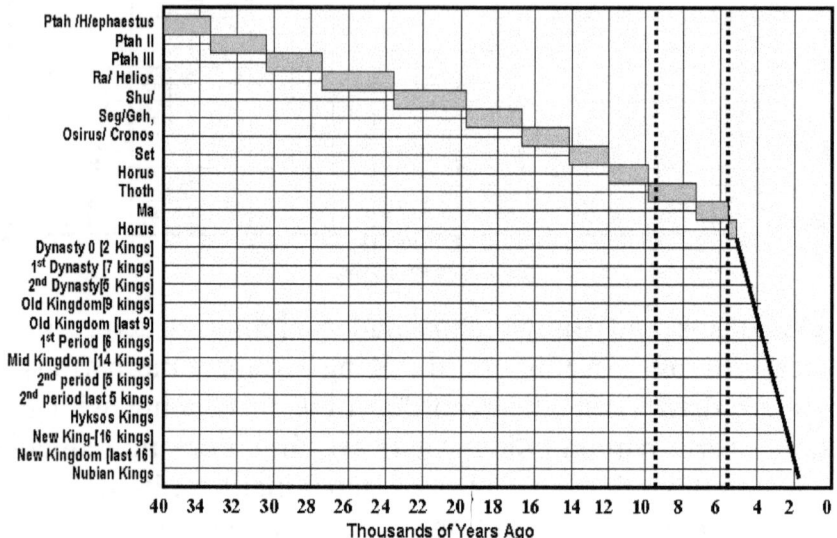

Thousands of Years Ago

I believe there is more than enough evidence to support the theory that when the Tower came down, people had shorter life times and were less intelligent. The shortened life spans are easily recognized. The intelligence limitation can also be recognized and tracked.

Jesus

It has taken us the last 6 thousand years to get back up to where we were before the Tower of Babel Wars and soon another we will be having another significant change affecting the earth. Some may wonder why I did not go over the God incarnate return time period 2 thousand years ago. While that element greatly affected human existence and evidence of his return is overwhelming, it did not generally affect the earth itself. Its significance is also not generally relative to this universe. What will affect the earth, in our universe, is the next Ice Age.

The Next Ice Age

Before I get into the next big change in our earth, let me review and expound on what we have already discussed.

Review

We have come a long way in determining how our earth developed. We found the following:

1. Several universes splatted together just before our current universe was made. This action set up the big bang and allow for galaxies to be clumped together because an intelligent designer controlled component called vibration.

2. The Big Bang splattered material all over the place about 15 billion years ago.

3. The solar system began coming together as some of the gases cooled in the Milky Way. By 4.5 Billion years ago the earth and most of the other planets were generally in positions associated with our solar system.

4. After at least 2 earlier encounters, Mar and the Earth came too close to one another and there was a catastrophe on both planets. The earth lost a large hunk of land where the Pacific Ocean is today and Mars lost about half of its surface. The earth's moon was also born.

5. The material expelled from the two planets started orbiting around the sun as asteroids during the Triassic Period.

6. The earth sped up for a time as its mass became less. This allowed large animals to exist and thrive.

7. At the end of the Cretaceous Period the earth slowed down when a large meteor struck the planet. The dinosaurs could not live with their new found weights and they died.

8. The earth's orbit was fairly close to that of its sister Venus during the years that followed the near collision with Mars that pushed earth closer to the sun. Venus looked like a huge wavy

haired planet in the sky. Some indicated it was as large as the moon.

9. Everything was fine on both Venus and Earth until something happened to the Venusian Moon. It shattered 11 or 12 thousand years ago and the destruction that followed was unparalleled.

10. On Venus, everything caught on fire, everyone died, the planet's spin slowed making the temperatures even hotter.

11. On earth, thousands upon thousands of meteors struck the planet, great earthquakes and a change in the rotational axis spelled doom for many. The temperature of the planet rose slightly and the current Ice Age came to a quick end. A third of all living things died and an island nation possibly named Atlantis sank into the Atlantic Ocean.

12. Three thousand years later, it is believed that a comet spelled doom for more animals and humans on the earth. The strike forced the earth's axis to change drastically and the poles melted which billowed water around the world. This was quickly followed by tidal waves unlike any witnessed in current times. Almost everyone and everything had died.

13. The survivors included geneticists who quickly remanufactured some of the lost animals and the earth quickly relived along with small patches of survivors around the world.

14. Two thousand years later, about 5000BC, people had regained control of the world this included worldwide use of electricity and unbelievable machines that used mercuric motors as their propulsion systems. Flying machines and massive weapons had been reengineered from designs used before the flood..

15. While we may not know the reasons behind the beginning of a terrible war, we do know the effects. We have found many signs of nuclear attacks, and massive destruction around the world. As a diversion of continuation, there began an assault into space or to the other "parallel" universe. While the exact details of the assault are not known, the outcome is very well known. Misery had again overtaken the earth.

16. About 1/3 of the people had again died in the battles and effects of the war. The remaining portions were not much better off. We are told that they became like apes. We are told they lost the ability to communicate. We are told that their "eyes had dimmed". Even the most impressive monument to human capability had been destroyed. That example was known as the Tower of Babel.

17. It is believed that a germ, probability a manufactured item of the horrible war, was set free. The germ didn't just kill. It modified the DNA of humans. They could no longer use all of their brains and the human brain began to atrophy. It has reduced in size by about 10% already when compared with an equivalent Neanderthal, prewar brain. Every year the brain is getting smaller as most of it is no longer used. People no longer remembered the sciences of the prewar people. They had to begin all over as if they were subhuman or ape-like.

After another 5 thousand years, we come to today. People are again gaining in intelligence, but even today our brains cannot be used properly. That doesn't really matter to the earth, but there is something we should look at as we investigate earth's changes . That element is our next Ice Age. Some have called it a time of Global Warming.

Global Warming

In a previous section, I used the Ice Core data to show when the worldwide flood occurred, but there is so much more that can be learned from ice coring. While on this subject, I just have to bring up the "Global Warming Theory". Let's see what the ice tells us. Many use Greenland graph [the >64N] graph below alone to show global warming but you can see it isn't the whole story as Antarctica is getting colder.

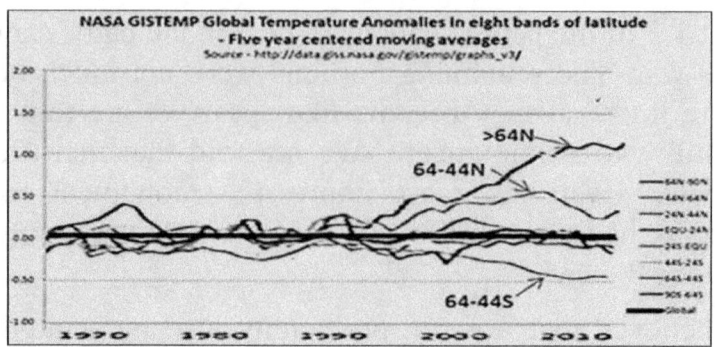

In the Greenland Ice Cores there is no indication of a major temperature increase except the high temperature spike with a 1.5 degree increase about a thousand years ago in the Greenland sample below. As shown above we have had about a 1 degree recovery, but it follows the standard thermal cycle of the chart below.

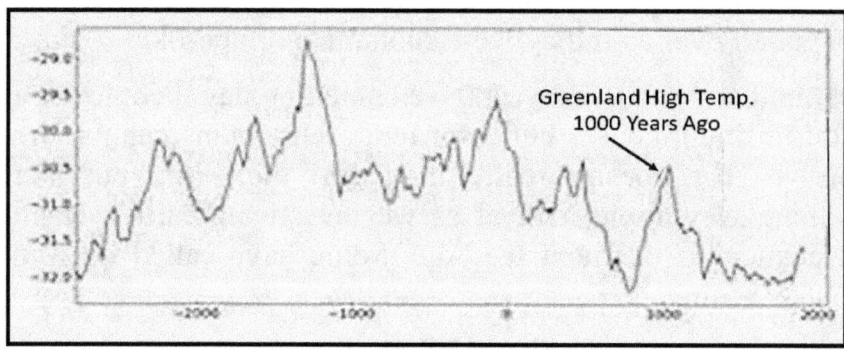

I know there were too many cars on the roads 1000 years ago, but most of the engines used Hay instead of gasoline. It looked like a major warming time without an equal cold period in Greenland and I don't think that that warming cycle occurred from too many people spraying deodorant or burning coal. Please don't let people say what arbitrary information means without showing you the data. Also remember that data can be easily misinterpreted so cross compare data from multiple sources. As an example; if there really was a worldwide flood 10 thousand years ago there should be evidence even in the Ice and if the earth were getting warmer we would see a trend. The graphs below are composites of a number of ice core temperature samples. If we are getting ready to "greenhouse", isn't the slope going the wrong way over the last 10 thousand years?

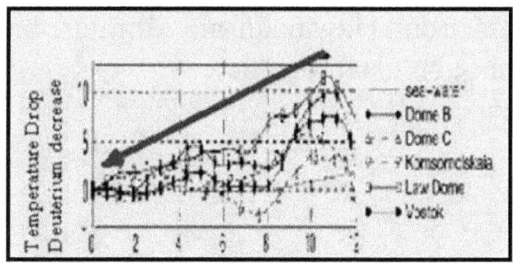 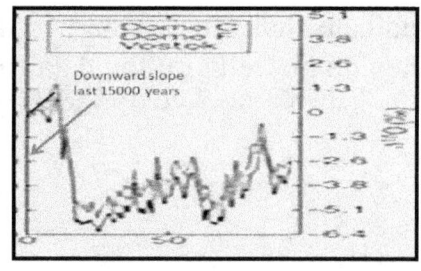

That being said, there has been a slight increase in effective temperature derived from CO2 data, but it has been greatly skewed. If we first look at the short term heating of the earth, the chart below shows that while the northern hemisphere is getting slightly warmer, the southern latitudes are getting colder. [Graph lines are only 0.5^0]

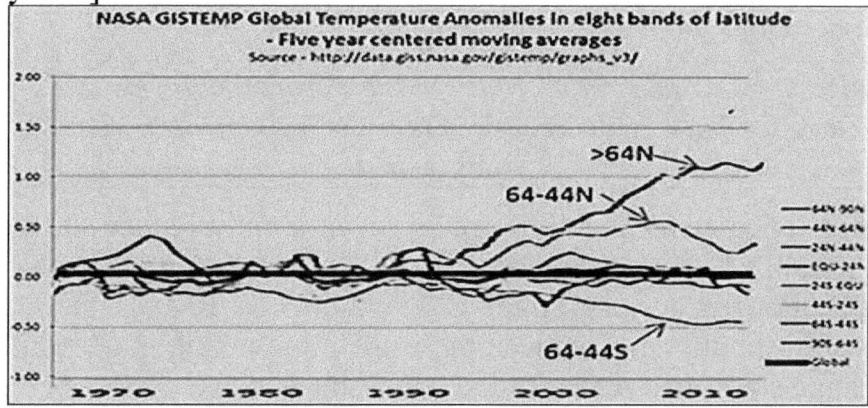

Its no wonder the predicted "flooding as the Oceans rise" scare has been finally halted with quasi-scientists saying the ground is sucking up the water. By the way please tell me you know that is another lie; which brings me to the main "INDICATOR" provided by NOAA. The Ice Core CO2 level graphs.

Ice Core Manipulation Dishonor

Here is the problem. Goverenmental agencies took Ice Core CO2 levels and added aerosol CO_2 readings to Ice Core data to make global warming look bad. To show they knew how dishonest this was it was <u>after</u> they published details indicating it takes hundreds or even thousands of years for Aeresol CO_2 to finally get absorbed by the ground. The <u>first graph below</u> shows the actual Ice Core data and the <u>second one shows the "sky is Falling" graph</u>. The graphs indicate that the last part is data from Hawaiaan air samples, but they do not tell you the data cannot be used together.

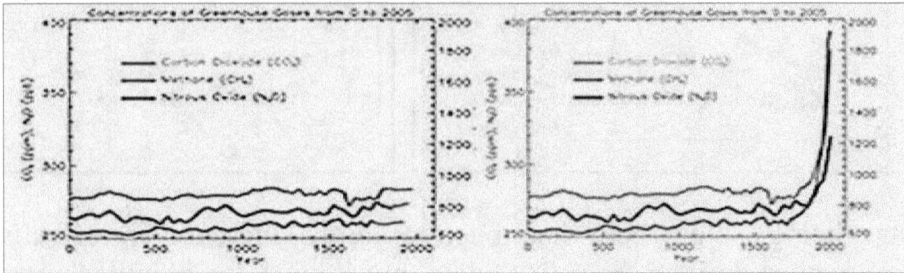

Let me show you how it was done. The first graph following shows the transition between CO_2 measurements in Antarctica ending in 1996 and Aeresol CO_2 measurements from Hawaii and from satelite imagry that started about 1960 and is shown with directed data up until 2008.

Notice there is a 33% difference in the absorption levels of CO_2 in Antarctica and the aeresol figures and the slope of increase is 3 times as fast.

Graph 2-Knowing the abosorption level, simply reduce the slope of Aeresol CO_2 by a 1/3 and what you have is shown in the middle graph showing a very slight CO_2 within "NORMAL" levels [I probably should have reduced the slope more taking in account for the huge slope differences of the 2, but this is worst case.

Graph 3-Instead of using logic, NOAA used trickery and misdirection to make the last graph that has been scaring everyone, including climatologists wanting there to be global warming to allow them to write papers and gain notoriety.

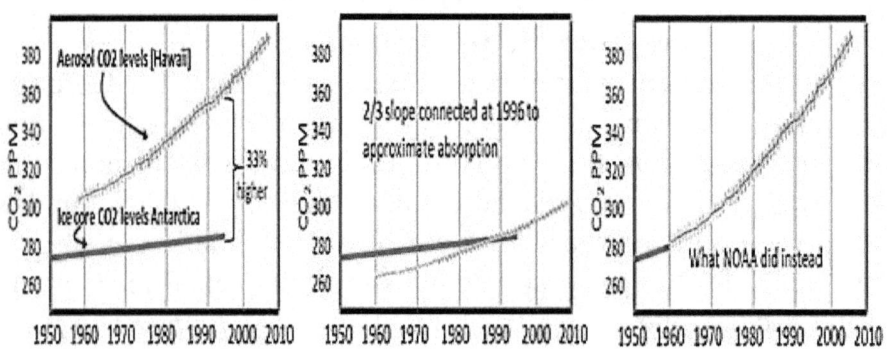

By putting the graphs together to make it look bad, they simply attached the Hawaiian levels onto the Ice Core data as if 100% of the CO_2 in the Hawaiian air gets to Antarctica and is not recombined before being buried in the Ice.

If we drill still further, the earth's cyclic thermal cycle is revealed and something very interesting. The graph below is backwards from the others so don't get confused. . By testing the ice cores we can get a possible thermal cycle of the earth. The data below shows that every 110 thousand years or so, the earth slowly gets colder and colder until the temperature bottoms out. This would presumably happen towards the end of an Ice Age. Only then does it appear that the earth can go through a significant heating cycle. The next heating cycle by this graph is about 80 thousand years in the future.   The ramp heights represent about 10 degrees in temperature. 10 degrees centigrade is the same 10 degree centigrade change noted in the Atlantic Ocean in our previous discussions about the sinking of Atlantis. The abrupt increase in Antarctica coincides with the increase in temperature identified in the Atlantic except it is shifted slightly. By about 3 thousand years which is the timing uncertainly level.

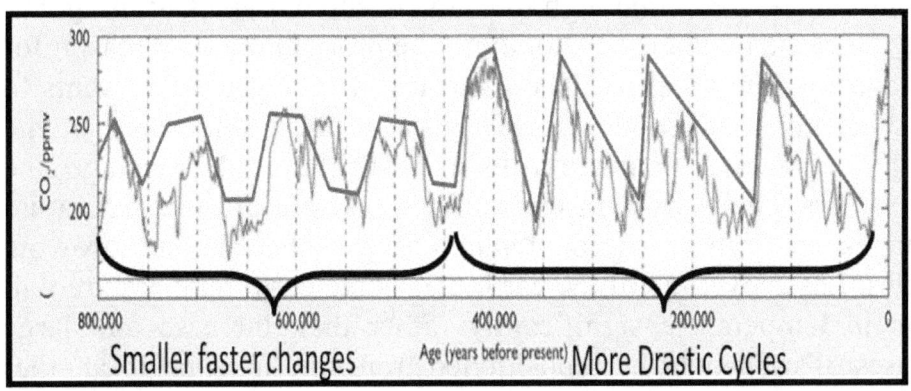

While we are at it, notice that about 450 thousand years ago the thermal cycle changes as if the size of our planet changed. Some would say that was when we lost the continent that is now the Pacific Ocean. I know I got carried away for a minute, but once you start looking at ice data it's like a time capsule just waiting to tell everyone about history. No one really knows why the ice temperature variances are as shown on the charts. While you have been told about the upcoming global warming incident, the whole thing is not supported by facts.

The Next Ice Age

Most likely, what is really happening, is just as bad as global warming, because we are evidently going steadily towards our next Ice Age. The ice core studies and long term weather research all indicate that the Earth alternates between ice ages and inter-glacials.

In 2002, at the South Pole, the penguins were stranded because of an abnormal ice buildup not because of the blazing heat the greenhouse model has suggested, but instead the weather was colder. This is only one of many indicators of the impending problem.

Destroying Trees

Scientists still don't know what even causes Ice Ages much less can they devise ways to slow its progress. We know that the bogus warning for people to halt cutting trees or the earth will get too warm didn't halt tree cutting. The idea of having too much CO_2 because there are not enough trees to consume it is a bad thing, but like the cattle methane, it should not be brought out as a reason for the earth getting warmer. The earth is still doing what it wants to do. One thing that does absolutely cause the homeostasis of the earth to change is a change in its axis of rotation. If a change like that occurs, plants won't grow as much because of the temperature changes and other environmental changes that disrupt growing cycles. Less coverage of the earth means the earth temperature will drop in temperature significantly rather than increase and large masses of ice would be repositioned from one location to another which would further amplify a delicate situation. This may be the way an Ice Age normally begins. By the way, I'm all for halting the extermination of our rain forest, because I like the medicines we get from them, and the oxygen they produce. Fewer trees is not good for any of us.

Pole Shift Catastrophe

Quite a few researchers are now indicating that our next destruction period will not come from the sky, nor will it come from war or even a simple Ice Age. Our next global catastrophe may be another one of those shifts in the rotational axis that caused the Mammoths to freeze about 10 thousand years ago. These scientists have cause to instigate concern as we look at the findings below that are not widely disseminated.

Magnetic Pole Wander

As I mentioned previously, the magnetic pole has wandered all over the place, and you may think that the wandering was only in ancient times, but we know how dramatic this movement is today because of the work of Paul Serson and Jack Clark, of the Dominion Observatory. The pole wanders daily in a roughly elliptical path around its average position, and may frequently have movements as much as 80 km away from this position when the Earth's magnetic field is disturbed. Accurate observations by Canadian government scientists in 1962, 1973, 1984, and most recently in 1994, showed that a northwesterly motion of the pole is continuing, and that during this century it has moved on average 10 km per year. The chart below is a running 30 day plot of the readings taken from one of several fluxgate magnetometer sensing sites placed around Canada and Alaska to check for movement of the earth's magnetic field.

- The component with larger spikes is positive magnetic northward

- The other component is positive eastward

Note the wild movement of the Earth's magnetic field as described by this instrument. First the field jumps to the Northwest then moves back eastward, followed by a southward travel and then switching east.

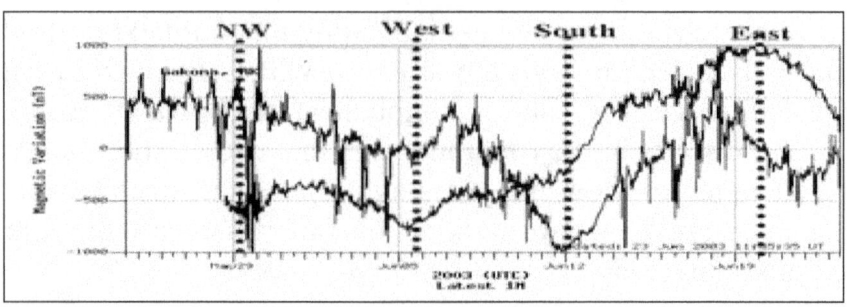

Our Polar Satellite data confirms magnetic movement. The graph below shows magnetic wander about the North Pole as captured from space. The general trend towards the northwest can easily be seen.

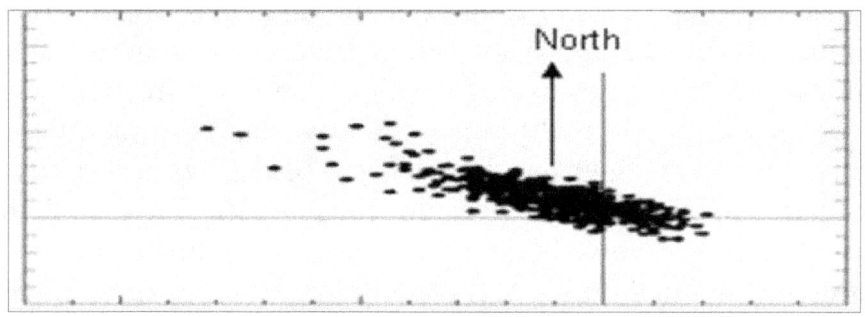

Scary Fluctuations

Since 1994 the average speed of the magnetic drift has **increased** to an average of 15 km per year which makes me sit up a little, but the story gets even more disturbing for those of us who would like to stay in the same climatic position. Satellite data has now been used to compare the strength and direction of the magnetic field in 1980 and again in 2000. According to researchers at "Physics of the Globe Institute" of Paris and the "Danish Space Research Institute", the magnetic field off the southern tip of Africa has already moved dramatically. I know that needs a little explaining. For that we go to Gauthier Hulot, a member of the Danish research team. He and his teammates indicated that molten iron under Africa is now moving in a direction, which will gradually weaken the dominant magnetic field and then reverse it. If the trend continues, the research shows that, we could be seeing the first steps toward a new North Pole. The timing has not been brought out, but saying "any time in the near future" is not good for me. I'm quite used to the hot Florida weather where I live.

Researchers have determined that the ultimate cause of the magnetic fluctuations is the Sun. The Sun constantly emits charged particles that, on encountering the Earth's magnetic field, cause electric currents to be produced in the upper atmosphere. These electric currents disturb the magnetic field, resulting in a temporary shift in the pole's position. The distance and speed of these displacements will, of course, depend on the nature of the disturbances in the magnetic field, but they are occurring constantly. With the earth being generally a sphere and most of the "Iron" component being in a liquid state, the magnetic element of the earth has no particular position it likes or doesn't like. Therefore, it could shift at any time. Whenever the magnetic pole shifts, the rotation axis shift will soon follow as the spinning creates a differential voltage that can only be eliminated by having both elements in the same plane. The shift does not have to be very significant for a destabilizing effect and "Ice Age" to occur so get out your long underwear.

Space Catastrophe

I'm not sure what will thrust us into our next Ice Age, but I'm fairly certain that it is coming just the same. One possibility has scared many and continues to scare those looking forward in earth's development. That is the probability of a near term huge meteor or Comet attack. Even After I presented to you the details of the repeated and multitude of meteor attacks and the horrible outcome of those things, you might just think that the possibility is remote that something like that will occur in your life time. Therefore, I have put together a few bits of evidence to make you a little more uncomfortable.

Upcoming Comet Catastrophe

For hundreds of years we have had many people claim to be able to foretell the future. This large group continues to expound on all the tragedies that are upcoming. Luckily most are ignored and as we try to get a better picture of what our future has in store for us, we don't have much choice but to rely heavily on information from some of these Seers provided that a track record of accuracy can be established. That typically means that a seer can't be used to foretell the future until he has been dead and a major portion of his prophecies have come true. We have such a collection and from those few, we may learn that a huge catastrophe is coming soon. By all accounts it will be an attack by a comet or very large meteor. If history repeats, there will be only one final outcome. The same thing will happen as before. Many will be killed.

Dr. Barry Warmkessel and others have provided us with very clear insight into at least a portion of what our future may bring by interpreting 15th century seers, the ancient texts, and one 20th century seer. That future involves comets and meteorites. The information collected and translated includes **what** will happen, **when** it will happen, and **where** it will happen. The following determination was made from prophesies from the Bible, Mother Shipton (1488–1561-England), and Nostradamus (1503–1566-

France), Edgar Cayce (1877-1945- USA), and many other stories and traditions. If we want to get a feel for the future, these sources are probably our best hope. Many of the stories concerning this eminent comet strike are integrated within a much more detailed account of our future. In those cases I will try to hold the predictions intact to allow sequential visibility of our world as perceived by the various prophets. To start this section, this brief overview comes from Edger Cayce, the Biblical book of Ezekiel, and the Pawnee Indian traditions.

Ezekiel's Comet

I'm not going to get into the accuracy of Biblical prophecy, because there are so many predictions that have come true, there can be little doubt about its predictive quality. Many Biblical prophecies tell of an impending comet disaster. This one is in the book of Ezekiel, but others will be provided later.

Ezekiel 38

*I will pour down torrents of rain and "GREAT STONES" with burning SULFER on him and on his troops and on many nations. At that time there will be **a great shaking** in the land of Israel.* [The region of Israel being the location of a major disaster and the idea that a great stone causes the damage are both similar to the other predictions that will be provided later.]

Edgar Cayce' Comet

Edgar Cayce is a reasonable prophet to use because of his accuracy. He made predictions starting in 1901 and many of those have come true. Edgar Cayce gave over 14,000 readings covering more than 10,000 topics. This included great accuracy concerning:

The beginning and ends of World War I
The beginning and ends of World War II
Details the German, Japanese alliance
Details of the German Holocaust
The American Depression and coming out of it
The Deaths of Roosevelt and Kennedy in office.

Cayce also provided insight concerning the effects of a comet strike. Additionally he predicted that the earth axis will shift, and a terrible war will come in the near future. Other "modern Seer" discussions about this comet will follow, but this one tells us a general time period and the effect of its coming, so he's a reasonable starting point.

"The Earth's axis will be shifted by the year 2001, "so that where there has been a frigid or semi-tropical climate, there will be a more tropical one, By this time, a new cycle would begin." [We know that the earth is wobbling much more now than in previous years. It seems to be ready for a shift to occur soon. By this prediction the earth shift will come before the Comet/meteor strike indicated by others. Although his prediction is off a little as we haven't experienced the big shift as of yet. I think we can at least say Cayce believed the shift would happen very soon.]

"The earth will be broken up in the western portion of America. The greater portion of Japan must go into the sea. The upper portion of Europe will be changed as in the twinkling of an eye. Land will appear off the east coast of America." [These could be the effect of a shift in the earth's axis, a comet, or both, but certainly some significant environmental disaster is predicted prior to a huge war that is in our future.]

"Portions of the now east coast of New York, or New York City itself, will in the main disappear. This will be another generation, though, here; while the southern portions of Carolina, Georgia -- these will disappear. This will be much sooner." [This indicates two separate flooding times. The first, I assume would come from a comet strike itself, while a second may come from the earth shift that follows.]

I especially hate the comment about the southern part of Georgia flooding because I live in Florida.

Pawnee Comet

There are many more similar prophecies of the destruction from a comet. When we take them all together, there can be little mistake. The same thing is going to happen that has happened in our distant past. We will be plunged into another Dark Age that will come from a Comet strike. The Pawnee said it simply.

218

Pawnee Indian Tradition- *Finally, the earth is to end by meteors. After this event, the sun and moon will darken.*

Our Last Major Meteorite Strike

No one seems to talk about this event, and it happened in a very remote section of the world, but it could have been very noticeable if it hit anywhere else in the world. The date was **December 9, 1997.** At 5:11 A.M., crews of three trawlers at widely separated sites off south Greenland reported "a blazing fireball that turned night into day." At a distance of over 60 miles away, the flash was compared to that from an atmospheric nuclear explosion. Seismic tremors also emanated from Greenland, so the impact of a large meteorite is almost certain. So far, no one has found the remains of the huge meteorite, but you have to recognize how very desolate and impossible that area is to search. Luckily our last major meteor hit Greenland and not Disneyland or people would easily accept the event.

The 1997 event was small in comparison to those that will happen in the near future. Before the really big strike there is probably going to be a somewhat smaller one and scientists know about it. The most likely candidate is about 2 Kilometers wide and it is classified as NT7. I previously discussed this nemesis, but I think it needs to be reviewed. It is scheduled to hit in 2016.

Besides the huge amount of tracked data that I presented previously, we can use two things that are a little more exotic to find out about upcoming meteoric disasters. The first thing is to know that history repeats itself. Many meteors and even comets have hit in the past and they will hit again. Although tried and true, that method has no timing accuracy or reasoning. The second method is by listening to more prophets or seers. Most who claim to foretell the future do not, but some are incredible.

Mother Shipton's Revelation

First let's look at this Mother Shipton character. Her real name was Ursula Sontheil and she is a reasonable prophet to use because of her accuracy. Her prophesies have had an extremely good record for coming true just like those of the more famous Nostradamus and Dr. Casey characters. Here is a short list of extremely detailed and accurately predicted historical elements that she wrote about. People thought that she was a witch and eventually killed her, but before they did, she had unveiled the future. [By the way she accurately predicted the method and timing of her own lynching. I think it is better not knowing what the future will bring, sometimes.]. She predicted the following:

Automobiles, the rise of the Church of England,
radios, telephones, telegraphs, hydroelectric power,
manufacture of mountain tunnels, submarines, airplanes,
Iron ships, the California gold rush., World War I
US Civil war and the French Revolution.
airborne military and their use, commercial air travel,
British and French alliance during the world war,.
The Allies and Communist bloc, and the cold war
The France to England underwater tunnel.
women would commonly wear pants and have short hair [an unthinkable thing at the time]
assemblies would be put together with huge machines,
the printing press and how it would change writing forever.

Shipton's Comet

With all of that as verification, I think we can use her other predictions of still more distant future events.

For those who live the century through- in fear and trembling this shall do. "Flee to the mountains and the dens-to bog and forest and wild fens. For storms will rage and oceans roar when Gabriel stands on sea and shore, and as he blows his wondrous horn old worlds die and new be born. **[This catastrophe will happen very soon.]**

*God's messenger from the heavens (**comet**) arrives and a great sound is heard as it passes through Earth's atmosphere and impacts Earth. It causes wild storms and raging seas.*

*A fiery dragon (**a comet as marked by it "dragon tail"**) will cross the sky six times before the earth shall die. Mankind will tremble and frightened be for the six heralds in this prophecy.* [It could mean six major meteor strikes will occur; spawned by the comet strike. It could also mean that before a terrible comet strike, we one earth will witness the huge ball of material getting closer and closer for 6 days before the eventual impact.]

For seven days and seven nights man will watch this awesome sight. The tides will rise beyond their ken. To bite away the shores and then the mountains will begin to roar and earthquakes split the plain to shore. [The strikes happen over a seven day period or on the 7th day.]

And flooding waters rushing in will flood the lands with such a din that mankind cowers in muddy fen and snarls about his fellow men. He bares his teeth and fights and kills and secrets food in secret hill and ugly in his fear, he lies to kill marauders, thieves and spies. [People begin killing each other over food.]

The world upside down shall be, and gold found at the root of a tree. [This is saying that the Earth axis will shift either before or after the comet strike. It is not known when.]

Yet greater sign there be to see as man nears latter century. Three sleeping mountains gather breath and spew out mud, and ice and earth and earthquakes swallow town and town. [Her centuries ended on the 26th year after a normal turn of the century, so she was talking about the time between 2001 and 2026. Earthquake and Volcanic action becomes significant. Both could be caused by the comet or the upcoming Earth axis shift.]

Nostradamus' Revelation

Old Nostradamus was no slouch. He complied hundreds of verses that appear to be very accurate predictions of his future. Many of these verses have come true as indicated below.

He predicted Hitler's reign and called him by the name of Hister.

He called out events associated with and correctly named Franco

He called out Mussolini's reign and alliance with Germany.

He indicated that twin towers would be attacked from the sky that occurred September 11, 2001 in New York.

He Named and dated the great fire of London.

He had many predictions concerning Napoleon which came true.

And many, many, many more

Here is what he had to say about the twin towers attack:

*A New City on the 45th latitude [**Guess where New York is!**] is attacked with fire from the sky as the North is put to the test. The earth shaking fire from the "World's Center" burns around the New City, two rocks make war on each other. The eagle-like attacker [**Jet**] of the New City is at first uncertain, then magnanimous victory, with damage to Cremona and Mantua [**These are the names of two giants like the twin tower giants**].*

Nostradamus Comet Attack

I mentioned that Nostradamus provides us with details and the comet event is a good place to start. Nostradamus tells us of a similar comet catastrophe to that of the preceding stories, but in this case he may have even told us where the meteors hit. His prophecy is so descriptive, we can build a map and track the invader. He tells us when it will hit, the size of the rock, and where the pieces will land.

Quatrain II:46- *After great human misery, a greater one approaches, The great motor of the **century renews**: It will rain* *blood, milk, famine, iron and pestilence. In the sky will be seen a fire, dragging long sparks.-* [A large meteor strikes a short time after the new century. They will introduce famine and disease.]

Quatrain I:69-*The great mountain, 4247 feet in circumference, After peace, warmth, famine, flood: The impact will spread far, drowning with great oscillations, even ancient objects and their great foundations.* [The diameter of the meteor is 400 meters. It causes great tsunamis. Even very ancient objects are destroyed.]

Quatrain VIII:16-*At the place where JASON had built his ship, There will be such great and so sudden a flood, That one will have nowhere on earth that was not attacked.* [Floods are felt around the world, but possibly Greece is hit first.]

Quatrain VI:6- *There will appear towards the seven stars. Not far from Cancer, the bearded star:* **Susa, Siena, Boeotia, Eretria, Great Rome** *will die, the night disappeared.* [Possibly indicates that the destroying comet will first appear near the constellation cancer. Again Mediterranean cities are mentioned on the destruction path.]

Quatrain III:10-*Greater calamity of blood and famine. Seven times it will advance toward the marine shore:* **Monaco**, *from hunger, place captured, captivity, the great leader is crushed in a metal cage.* [The comet splits into 7 pieces before hitting along the shoreline near Monaco.]

Quatrain I:46-*Very near to* **Auch, Lectoure and Mirande**, *Great fire of the Sky in three nights will fall; which will cause a stupendous event to happen. A short while after the Earth will tremble.* [This verse indicates that one meteor strikes near the three French towns mentioned. The strike initiates huge earthquakes.]

Quatrain V:98-*At a* **latitude of 48 degrees** *climatic. At the end of Cancer so great is the drought: Fish in the sea, river and lake boiled hectic, Bearn, Bigorre the sky in distress from fire,* [In western France, a great drought is felt from the fire in the sky.]

Comet Strike Plot

From the above predictions and the interpretive work of researcher Barry Warmkessel, we can make out the destruction path. The picture below shows the intersection of the prophesized hits- Susa -

Italy/France Border, Siena - Tuscany Italy, Eretria – Greece, Monaco, Boeotia – Greece, Mount of Olives- Jerusalem, Rome-Italy, Auch, Lectoure, Mirande, Bearn, Bigorre-France, and Sardinia.

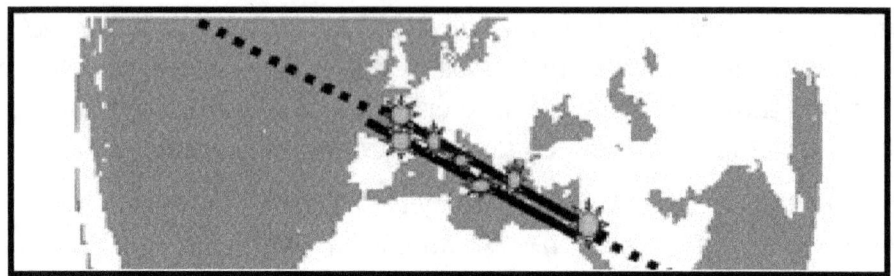

Comet Timing

In addition to the few just presented, predictions of this meteoric disaster occur around the world. Some of the predictions even told us when. The following is a mixture of ancient writings and scientific discussions. We don't know exactly when all the predictions will come about, nor can we be assured that they will, but we may have a better idea about the beginning of the end if we try to cross-compare "prophecies". By these insights, the beginning will be the comet strike. Here is what has been told around the world. Look for the similarities of the impact event and when to expect the catastrophe.

South American Prophecy

Incan Prophesy- The end of the age of man will be a _Rain of Fire-_ **about 2012.**

North American Prophecy

Hopi/ Apache Prophesy- The end of the age of man will be _Destruction by Fire around_ **2000.**

Sioux Indians Prophesy- The end of the age of man will be the era of the _pipe [fire]._

Central American Prophecy

Aztec Prophesy- The end of the age of man will be Destruction by blood and _fire_.

Toltec Prophesy The end of the age of man will be _Destruction by devastating wind_.

Mayan Prophesy- End of the world according to the Maya, or at least, the end of their calendar is **December 23, 2012.**

15ᵗʰ Century Seers

Mother Shipton Prophesy- She indicated that the end will be in **2026**. -_For those who live the century through [the century for her_ *began in 1926]- in fear and trembling this shall do. "Flee to the mountains and the dens-to bog and forest and wild fens. For storms will rage and oceans roar when Gabriel stands on sea and shore, and as he blows his wondrous horn old worlds die and new be born.* [This catastrophe will happen just before the year 2026]

Nostradamus- *[III-94] For 500 years more one will keep count of him who was the ornament of his time. Then suddenly a great light will be given. "He who is for this century" [Nostradamus] will render them satisfied.* [This strange verse apparently tells us the great light of the comet is coming about 500 years after Nostradamus or in the 21st century.]

20th Century Seers

Rubinsky- This 20th century seer indicates that an environmental crash will occur **in 2100.**

Edger Cayce Prophesy- As we previously discussed, the twentieth century seer, Cayce, stated in 1934, "there will be a shifting of the poles **by the year 2001.** Upheavals in the Arctic and the Antarctic will make eruption of volcanoes in torrid areas. The Earth will be broken up in the western portion of America".

Astronomical Events

If we go back to science for a minute, here are the dates of importance.

The Comet Wirtanen- This comet is due to pass within less than 1 AU of the Earth in the year **2013.** The comet itself is about twice the size of Halley's comet and could do so pretty terrible damage if it hits.

Other Meteors- If the comet doesn't hit, there are 2 other major asteroids poised for an impact. The first is named "WN5 and it will pass within 0.00015 AUs [125 thousand miles] in the year **2039.** That is followed by one named "WO107 which passes in 2140 at a distance of 0.0005 Au's [plus or minus 0.01 Aus.]

Photon Belt- According to Astronomers, there is a charged area in the universe, which our solar system periodically enters. Whenever we enter much change occurs. The next entry date is supposed to be **2012.**

Scientific American- Magnetic field reversed more than 170 times in past 80 million years last reversal [using the old nuclear decay dating], according to "Nature and New Scientist" and "Scientific American" magazines, occurred 13 thousand years ago and the **next shift is expected around 2030AD.**

Biblical Interpretation

Daniel was a great prophet depicted in the Bible. We may glean some details about the coming of this "comet" from his predictions.

Daniel 12:7- *"It shall be for a time, times and a half:* ***[2.5 time periods] (If a time period is 1 thousand years it would be 2500 years or*** *about the year 2000.) and when they have made an end breaking in pieces the power of the holy people, all these things shall be finished-* [This verse may indicate that there will be 2,500 years from the time of Daniel until the comet attack.]

Daniel 12:11- *from the time that the burnt offering shall be taken away [time of Jesus] and the "abomination that makes desolate" [possibly the Moslem rule] is set up, there shall be 1,290 time periods [years] Blessed is he that waiteth and cometh to the 1,335 time periods [years], but go thy way until the end of the days."*

If we assume the time periods are years, it seems to indicate that 1290 years after Jesus was executed, a large group against the teachings of Jesus arose and will be influential until the year 2325. If this group is the Moslem nation, then the 1290 AD date indicates the time when the Mongols had taken over much of the Middle and Far Eastern world and had adopted the Moslem religion. This made that "group", the huge power it is today. If this is the meaning, then a major war that occurs after the comet attack will be ended by about 2325AD.

Conclusions

Hopefully you are at least more aware of the things that are around you and the limitations that our schools have when trying to force a comfortable earth history down our throats. If you don't take anything else away from this book please take away the concept that for religion and science to be correct both must work together. Also remember that science, without religion has not been able to explain the developments on our earth. Before I close, let's review some of the things we have found out about the development of the earth, hopefully your insight has been broadened or your beliefs have changed.--- Now you have more data to determine a reasonable truth.

Big Bang Theories are all flawed, but can come together with a "creator" sort of driving the boat.

The current String Theories do something very important. They REQUIRE the existence of a parallel universe like "heaven" noted in religions around the world.

Atomic Theories don't cut it anymore. There is a unified Particle that everything is made from and that particle is called a BOSON.

Vibration-controls invisibility and matter.

Einstein's Theory of Relativity- Is in error.

Matter- About ½ of all matter has no mass

Heavy Dinosaurs- They were not.

Mars- made most of our mountains and the Pacific Ocean. Finally, ½ the surface of Mars was ripped away.

Venus- A mere 12 thousand years ago, it almost split in half, its moon exploded, parts of the Venusian Moon pelted the earth, and the destruction started the whole planet on fire.

Planets- were probably visited by humans many years ago.

Ancient humans lived over a million years ago. They were not idiot ape-men, but instead were highly civilized.

The Flood- It really happened, but tidal waves killed the population.

The Tower of Babel- was a symbol of one of the most devastating wars on earth. Mankind was changed by some horrible germ that was freed during the war.

Electricity- The great pyramid looks more like an electric generator than just about anything else and ancient humans used electricity. Both seem to go together.

The next Meteor- is coming soon.

The next Ice Age- is nearly here.

www.ingramcontent.com/pod-product-compliance
Lightning Source LLC
Chambersburg PA
CBHW051902170526
45168CB00001B/207